Autorenkollektiv

Small Ring Compounds in Organic Synthesis I

AKADEMIE-VERLAG BERLIN

Autorenkollektiv

Small Ring Compounds in Organic Synthesis I

Autorenkollektiv

Small Ring Compounds in Organic Synthesis I

With 1 Figure and 27 Tables

AKADEMIE-VERLAG BERLIN
1986

Die Originalausgabe erscheint im Springer-Verlag Berlin · Heidelberg · New York · Tokyo
in der Schriftenreihe „Topics in Current Chemistry", Volume 133

Vertrieb ausschließlich für die DDR und die sozialistischen Länder

ISBN 3-540-16307-7 Springer-Verlag
Berlin · Heidelberg · New York · Tokyo
ISBN 0-387-16307-7 Springer-Verlag
New York · Heidelberg · Berlin · Tokyo

ISBN 3-05-500202-4

Erschienen im Akademie-Verlag Berlin, DDR-1086 Berlin, Leipziger Straße 3—4
Lizenznummer: 202 · 100/534/86
Printed in the German Democratic Republic
Gesamtherstellung: VEB Druckerei „Thomas Müntzer", 5820 Bad Langensalza
LSV 1275
Bestellnummer: 763 638 0 (3072/133)
12400

Guest Editor of this volume:

Prof. Dr. *Armin de Meijere*, Universität Hamburg, Institut für Organische Chemie und Biochemie, Martin-Luther-King-Platz 6, D-2000 Hamburg 13

Table of Contents

Introduction

Small ring compounds have come of age. The year 1985 marked the 100th anniversary of Adolf von Baeyer's "theory of ring strain" (published in 1885) as well as his 150th birthday. Now, a handsome one hundred years after the first cyclopropane and cyclobutane derivatives were synthesized and their propensity for ring opening discovered by William Henry Perkin, synthetic organic chemists have begun to realize on a wide scope that there is more to these compounds than fun and intellectual exercise for the structurally oriented chemist.

The past 30 years have seen the development of a broad spectrum of widely applicable preparative methods for three- and four-membered carbocycles and the accumulation of detailed knowledge about their structure-reactivity relationships. Nowadays, more and more synthetic methodology is being developed, which utilizes the potential of small ring compounds as reactive entities. Cyclopropyl and cyclobutyl moieties in a molecule can be regarded as unique functional groups; they allow transformations which are far more difficult or impossible with any of the more conventional functional groups.

To the extent that the bonding in a three-membered ring resembles that in a C=C double bond, most of the chemistry of cyclopropane derivatives can be rationalized by analogy to similarly substituted olefins. By their electronic nature, therefore, cyclopropanes are nucleophilic, that is to say their high reactivity especially towards electrophilic agents is not solely governed by ring strain. Furthermore, the philicity of a cyclopropane can be modified in any desired way by substituents. Like a double bond, it can be made susceptible to nucleophilic attack (e.g. the Michael addition) with strongly electron withdrawing substituents; practically then, cyclopropyl ketones are "homoenones". Electron donating groups increase the nucleophilicity drastically; consequently cyclopropyl ethers behave as "homoenolethers", cyclopropanolates as "homoenolates". In all cases the cyclopropyl homologues of the corresponding olefins always lead to a substitution pattern after ring opening which could otherwise only be achieved by way of an "Umpolung". In essence, the use of a cyclic array of three carbons is one way of performing an "Umpolung" at the γ-carbon of a three carbon chain and has therefore been termed the "cyclopropane trick" [1].

The combination of a donor and an acceptor on one cyclopropane ring creates yet another potentially useful reaction path. Finally, cyclopropanes may also be substituted in such a way that they are most easily opened by radical attack.

Vinylcyclopropanes are now widely used as precursors to five-membered rings. Various methods have been designed to construct vinylcyclopropanes and bring about their rearrangement — most frequently thermolytically — to cyclopentenes.

Some rather general ways of cyclopentene-anellations are based on this scheme. Substituents effects once again have been used successfully to facilitate the rearrangement, which — in the extreme case — occurs at room temperature. More recently, alkinylcyclopropanes have found interesting applicability as building blocks for organic synthesis, as have methylenecyclopropanes and cyclopropenes. Transition metal chemistry is starting to play a major role in this field, since many of the reactions of small rings can also be achieved with the help of catalysts. More work needs to be done in this area and some is already in progress.

The cyclobutyl moiety has also come to be quite frequently used as an all carbon functional group in efforts directed toward the synthesis of complex organic molecules. Four-membered ring chemistry can likewise be tuned by the appropriate choice of substituents. In addition to various chemoselective ring openings there are ring enlargements to five- and ring contractions to three-membered carbocycles, all of which have their specific range of applications.

This volume of "Topics in Current Chemistry" and another one, projected to appear within the next year are intended to cover most of the above-mentioned aspects of small ring chemistry, with emphasis being placed on the application in organic synthesis. "State of the art" reviews will be presented by individual authors who are themselves engaged in research on one or more of the subtopics. This exciting field has become so active that no single author would feel competent to write a comprehensive, up-to-date monograph.

Armin de Meijere

Reference

1. Cf. D. Seebach, *Angew. Chem.* 91, 259–278 (1979); *Angew. Chem. Int. Ed. Engl.* 18, 239 (1979).

Strain and Reactivity: Partners for Selective Synthesis

Barry M. Trost

McElvain Laboratories of Organic Chemistry
Department of Chemistry, University of Wisconsin, 1101 University Avenue Madison, WI 53706/USA

The unusual bonding of cyclopropanes and the strain release associated with cleavage of cyclopropanes and cyclobutanes offer the possibility of recognizing that structural unit when it is a part of a larger molecule. These structural fragments may be considered as "pseudofunctional groups" which may be chemoselectively manipulated. The structural and reactivity concepts have evolved into the development of reagents that incorporate a cyclopropyl ring into substrates to permit the development of chain extensions and annulations.

The attachment of a cyclopropyl ring to a carbonyl group requires the development of cyclopropyl anions. Of special interest are those bearing a heteroatom substitutent either at the anionic carbon or an adjacent one. The ability of sulfur to stabilize an anionic center has proved to be particularly beneficial for generating reagents of the first type. Two reagents l-lithiocyclopropylphenyl sulfide and diphenylsulfonium cyclopropylide generate a great flexibility for structural variation. Mimics of these reagents bearing oxygen and selenium substituents in place of the sulfur substituent have also evolved.

An alternative strategy invokes a need for an electrophilic cyclopropyl reagent to introduce the small ring at a carbon alpha to a carbonyl group via the corresponding enol or enolate. Two reagents belonging to this class, l-phenylthiocyclopropane-l-carboxyaldehyde and l-tetrahydropyranoxycyclopropane-l-carboxyaldehyde, utilize an aldol reaction to incorporate the small ring conjunctive reagent into the organic molecule.

Two major types of reactions of the small ring systems provide the greatest versatility. Ring enlargement to cyclobutyl systems creates another class of strained ring compounds that permit selective structural variation. Ring cleavage exemplified by the secosulfenylation of the cyclobutanones illustrate the development of a diastereoselective geminal alkylation of a carbonyl group. Further ring enlargement of the cyclobutyl ring to a five membered ring provides syntheses of both carbocycles, i.e. cyclopentanones, and heterocycles, i.e. γ-butyrolactones. An alternative provides vinylcyclopropanes and vinylcyclobutanes which can be induced to rearrange to cyclopentenes and cyclohexenes respectively. Cyclopentanone and cyclohexanone annulation result from these reactions.

The versatility of cyclopropyl conjunctive reagents permits the development of synthetic strategy to many types of complex molecules. The concepts of secoalkylation-geminal alkylation produce total syntheses of methyl deoxypodocarpate, hinesol, grandisol, trihydroxydecipiadiene, acorenone B and a verrucarol intermediate. Lactone annulation and substitutive spiroannulation create the basis for synthesis of plumericin, allamcin, allamandin, a verrucarol intermediate, and dodecahedrane. Cyclopentyl syntheses not involving γ-butyrolactones as intermediates lead to synthese of α-cuparenone, aphidicolin, spirovetivane, α-vetispirene, 11-deoxyprostaglandin E, and isoretronecarrol. A synthesis of β-selinene exemplifies the vinylcyclobutane rearrangement. A synthesis of methyl trisporate B illustrates the use of these small ring conjunctive reagents for formation of *acyclic chains*.

1 Introduction

The ability to perform selective transformations at one point in a multifaceted molecule requires a distinctive structural feature that can be recognized either due to its intrinsic reactivity or by the nature of the reagent or the catalyst. The latter area represents a major endeavor — especially with respect to the concept of enzyme modelling. For example, selective oxidation of a C—H bond at various "unactivated" positions in a steroid may mimic oxidases but also have practical synthetic uses. However, the former area, in one sense, is the more classical. The concept of functional groups which mean π (carbonyl, olefin, arene, etc.) or heteroatomic (oxygen, sulfur, nitrogen, etc.) systems formalizes this thinking. For example, deeply imbedded into the thinking of organic chemistry is the widespread utility of a carbonyl group for further structural elaboration. Regardless of how many other types of bonds that may be present in a molecule, we can manipulate around the carbonyl group quite selectively. While we have a well recognized list of such functional groups, the question arises as to whether the synthetic chemist can add to this list.

As an outgrowth of the extensive physical and theoretical studies of strained rings, the extension of the functional group concept to three and four membered rings becomes attractive. The unique bonding of cyclopropanes combined with the release of the strain of both the cyclopropane (117 kJ/mole, 28 Kcal/mole [1] and the cyclobutane (109 kJ/mole, 26 Kcal/mole) [1] rings provide a recognition mechanism for unique molecular reorganization of that portion of the molecule. To realize the benefits of such a "pseudo-functional group" we must develop ways for their creation or incorporation.

The discovery of carbene and carbenoid additions to olefins was the major breakthrough that initiated the tapping of this structural resource for synthetic purposes. Even so, designed applications of cyclopropane chemistry in total syntheses remain limited. Most revolve around electrophilic type reactions such as acid induced ring opening or solvolysis of cyclopropyl carbinyl alcohol derivatives. One notable application apart from these electrophilic reactions is the excellent synthesis of allenes from dibromocyclopropanes [2].

An alternative mode for harvesting this resource is to design conjunctive reagents that will directly introduce this structural feature into organic molecules. The widespread presence of a carbonyl group in organic molecules make it a target for attachment of the strained ring. Two types of reactivity derives from a carbonyl group — nucleophilic addition to the carbonyl carbon atom and electrophilic substitution at the α-carbon through the intermediacy of enols and enolates. For the first type of reactivity, cyclopropyl anions are required; for the latter, a cyclopropyl ring bearing a carbonyl group or another electrophilic functional group are desired. Of course, these types of reagents will not be limited to combining with carbonyl groups as some of the examples will demonstrate. For most cases, the cyclopropyl product is not the final goal, but an intermediate to acyclic or ring systems commonly needed in theoretical and natural products chemistry. For such applications, the parent ring system is less useful than those bearing additional substituents, especially heteroatomic groupings.

In designing conjunctive cyclopropyl reagents, we should consider some of the types of structural modifications that appear most useful for further synthetic goals. Such considerations define the nature of the substituents on the cyclopropyl ring and the type of reaction to be utilized.

2 Ring Opening

The cleavage of the cyclopropyl ring with its relase of the total strain is a powerful driving force. Electrophilic attack on the electron rich ring does provide one approach as shown in Eq. 1 [3]. The lack of selectivity in the cleavage of one of the

(1)

three cyclopropyl bonds usually makes it useful to incorporate directing substituents such as oxygen (Eq. 2) [4].

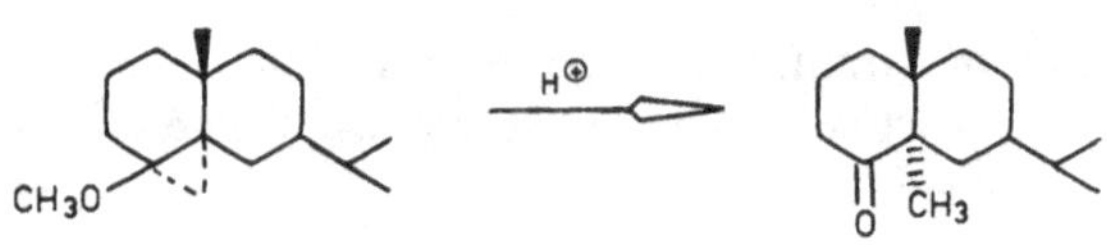

(2)

A cyclopropylcarbinyl cation can be trapped to form either a cyclopropane product (Path a, Eq. 3) or a homoallyl product (Path b, Eq. 3). The latter has proven useful to create acyclic units containing olefins of defined geometry as in the synthesis

(3)

of juvenile hormones (Eq. 4) [5].

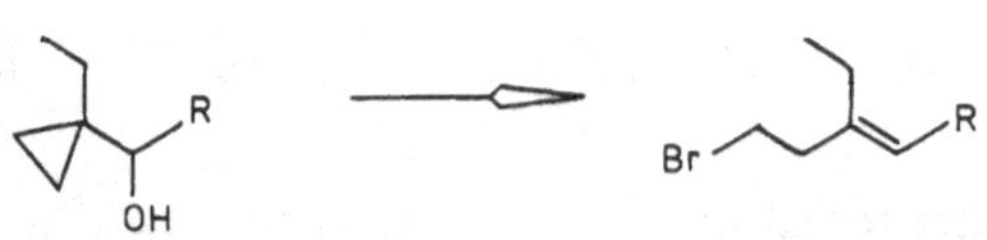

(4)

3 Ring Expansions

The unusual nature of the cyclopropyl carbinyl cation allows yet another mode of attack to form cyclobutane products. Because this mode of attack releases little strain, normally some special structural features are required to direct the reaction along this pathway.

(5)

The utility of this pathway mainly derives from the further reactions of the cyclobutanes. Since they possess nearly as much strain as the cyclopropanes, powerful driving forces for further structural modification still exist. For example, ring expansion to cyclopentyl rings or γ-butyrolactones (Eq. 6 [6,7] and 7 [8])

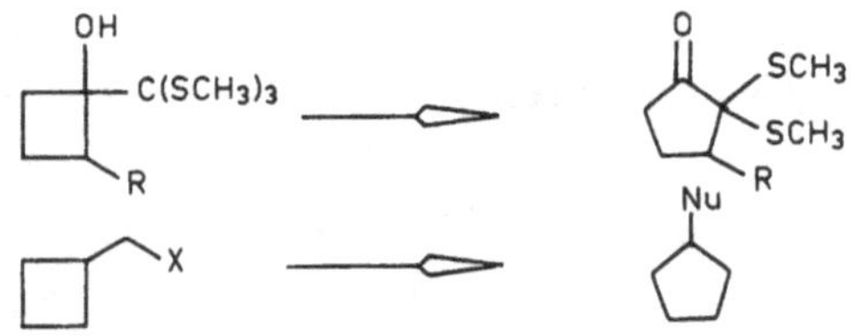

(6)

(7)

provide dramatic release of strain. Cleavage of cyclobutanones bearing anion stabilizing groups at the α-carbon atom has served as a stereocontrolled olefin

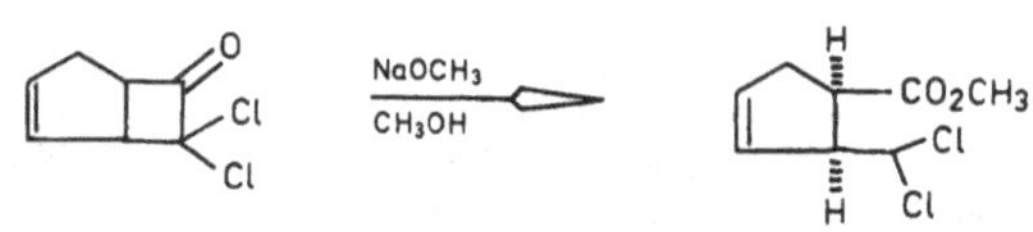

(8)

vicinal alkylation since the starting dichlorocyclobutanone or related systems can derive from ketene cycloaddition [9,10]. Since any cyclobutanone can be elaborated in such fashion, ring expansions to cyclobutanones become particularly valuable.

Vinylcyclopropanes represent particularly useful functionality. They do permit a ring expansion to cyclobutanes via the cyclopropylcarbinyl cation manifold (Eq. 9). Equally important, such systems suffer smooth thermal rearrangement to cyclopen-

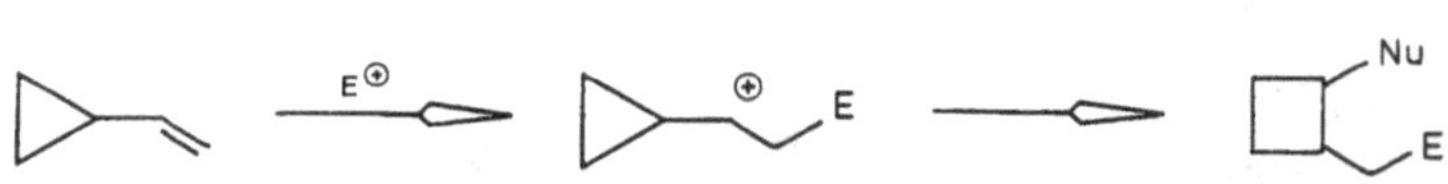

(9)

tenes (Eq. 10) [11]. Rate studies reveal that substitution at the one (Eq. 11) or two

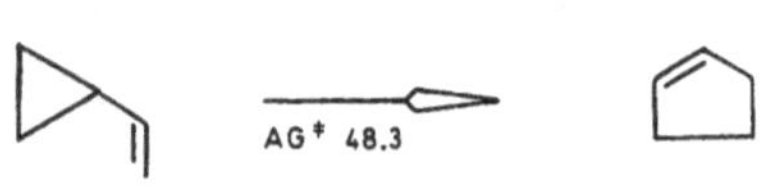

(10)

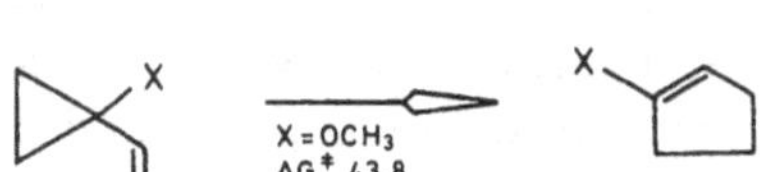

(11)

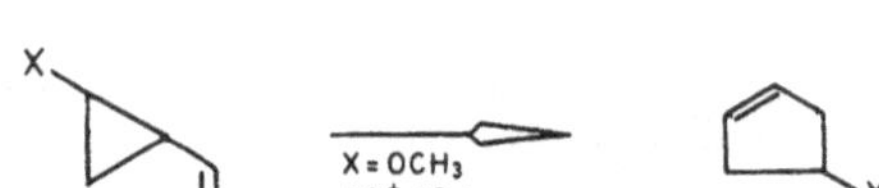

(12)

(Eq. 12) position of the cyclopropyl ring dramatically lowers the activation energy of this process [12]. In addition charge accelerated vinylcyclopropane rearrangements have been noted (Eq. 13) [13]. The structural flexibility offered by substituents

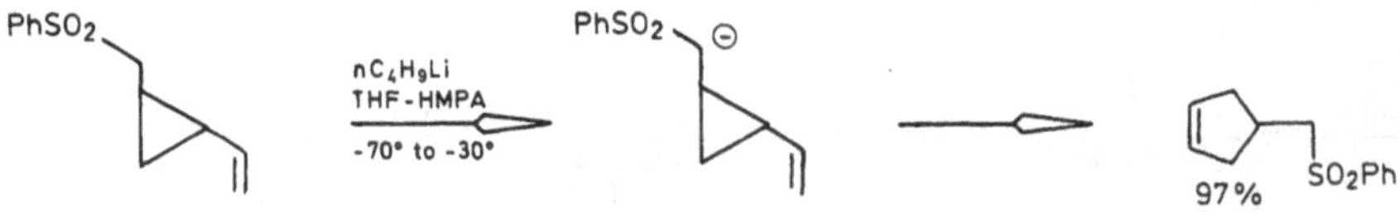

(13)

allows us to take maximum advantage of these rate accelerations.

Vinylcyclopropanes bearing a cis alkyl substituent undergo a competitive prototropic shift accompanying ring opening (Eq. 14) [14]. In such cases, temperature

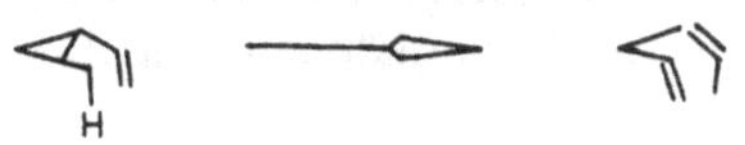

(14)

adjustment permits either pathway, i.e. cyclopentene or 1,4-diene formation, to dominate. Higher temperatures (> 650 °C) generally favor cyclopentene formation [15]. The presence of a β-vinyl substituent permits yet another pathway to intervene — the divinylcyclopropane rearrangement to cycloheptadienes (Eq. 15) [16]. While

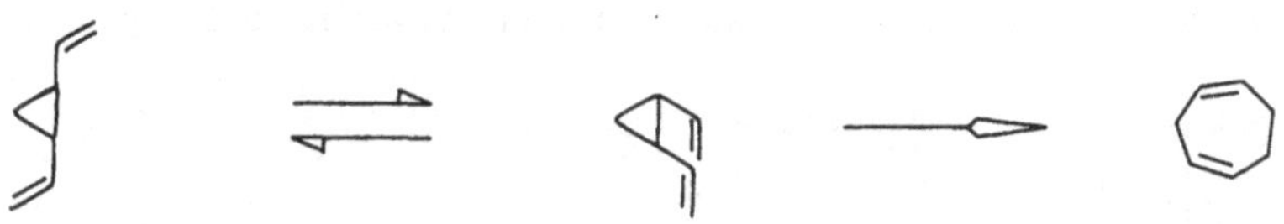

(15a)

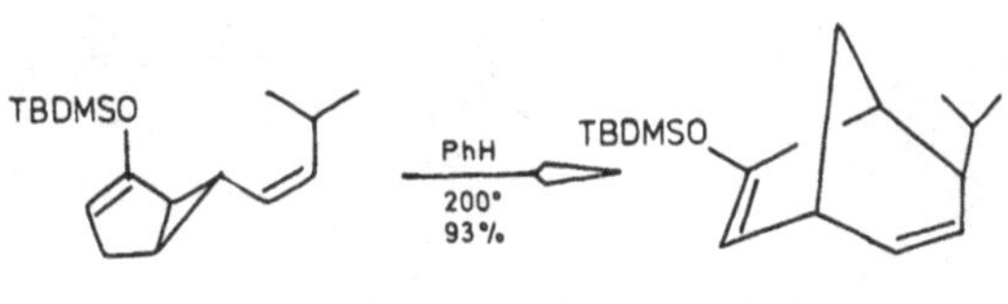

(15b)

only the cis isomer can undergo this reaction, the easy trans-cis isomerization under the reaction conditions allows both isomers to be used. Thus, manipulation of cyclopropyl substituents can provide diverse opportunities for a wide array of structural variations.

4 Cyclopropyl Anions

While thermodynamically, the direct metalation of cyclopropane can be envisioned from a synthetic point of view, this approach has been rarely used. A major obstacle appears to be kinetics which can be overcome by incorporation of a hydroxyl group (see Eq. 16) [17]. In special cases, such as bicyclo [1.1.0] butane and methylenecyclopropane (Eq. 17) [18] the enhanced thermodynamic acidity is accom-

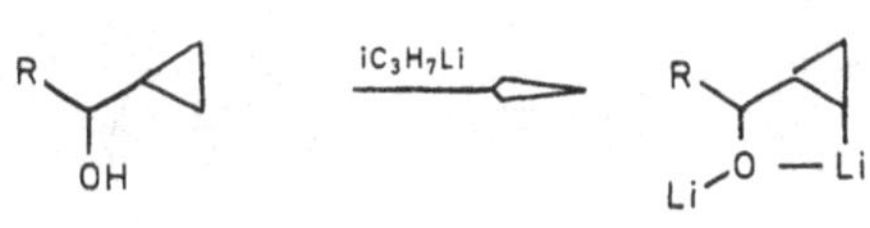

(16)

panied by an enhanced kinetic acidity as well. In the latter case, the anion adds to the carbonyl group of aldehydes and ketones. When the carbonyl partner is an

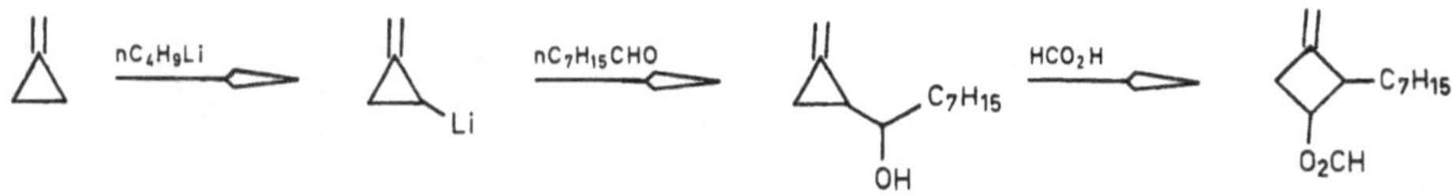

(17)

aldehyde, such as n-octanal, solvolysis leads to smooth ring expansion to give a 3-methylenecyclobutan-1-ol, the product of exclusive migration of the vinyl cyclopropyl carbon. Thus, even such simple systems translate into a useful cyclobutanol synthesis.

A more general approach utilizes metal-halogen exchanges because of the ready availability of monobromocyclopropanes by either
1) reductive monodehalogenation of dibromocyclopropanes or
2) halodecarboxylation (Hunsdiecker reaction)
of cyclopropane-carboxylic acids. For example, the parent cyclopropyllithium, available from bromocyclopropane, forms a cuprate for conjugate addition (*1* → *2*, Eq. 18) [19]. The resultant vinylcyclopropane can suffer a smooth molecular reorganiza-

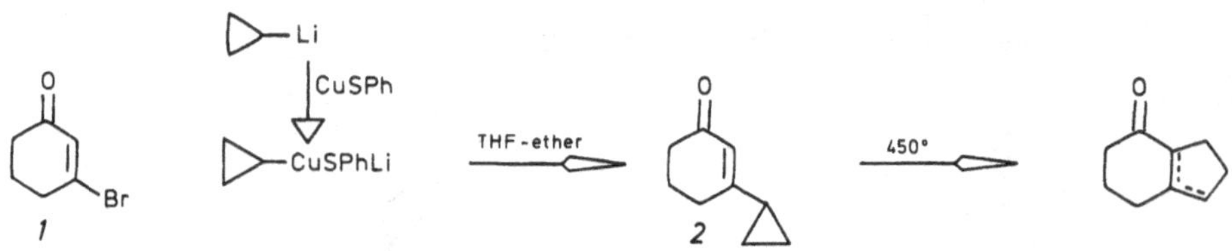

(18)

tion upon thermolysis to create a cyclopentannulation.

A slight modification of the cyclopropyl conjunctive reagent transforms a cyclopentannulation into a cycloheptannulation. Thus, the 2-vinylcyclopropyllithium reagent *3*, converted to its cuprate *4*, generates a 1,2-divinylcyclopropane. Heating to only 180 °C leads to smooth Cope type rearrangement, driven by the release of the cyclopropyl strain, to create a perhydroazulene ring system of many sesquiterpenoids (Eq. 19) [20].

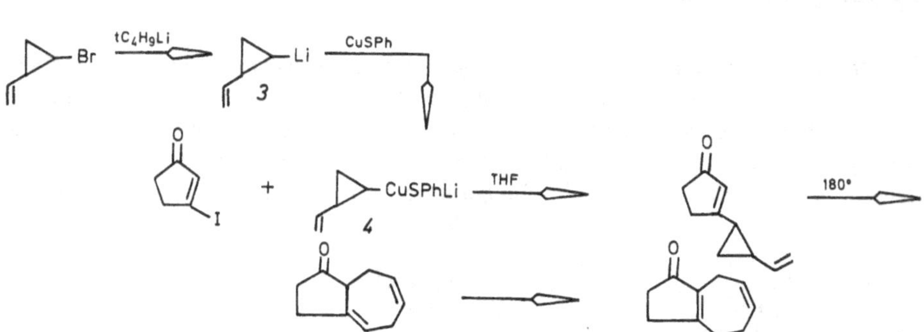

(19)

The dibromocyclopropanes offer an easy entry into alkylated versions of cyclopropyl anions. For example, the methylated cyclopropyllithium reagent *5* is readily available by sequential metal-halogen exchange, quenching with methyl iodide and metal-halogen exchange (Eq. 20) [21]. A divinylcyclopropane can be easily generated by addition to a carbonyl group of a β-methoxyenone as in Eq. 20. A cycloheptannulation results upon thermolysis in a complementary process to the conjugate addition approach of Eq. 19. A synthesis of Dewar heptalene employed this approach to create the requisite carbon skeleton (Eq. 21) [22].

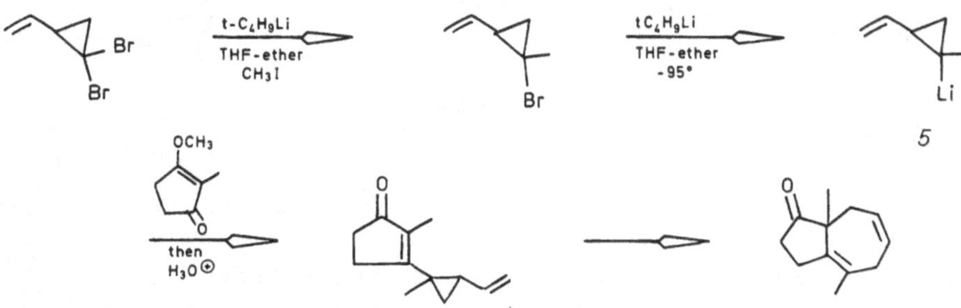

(20)

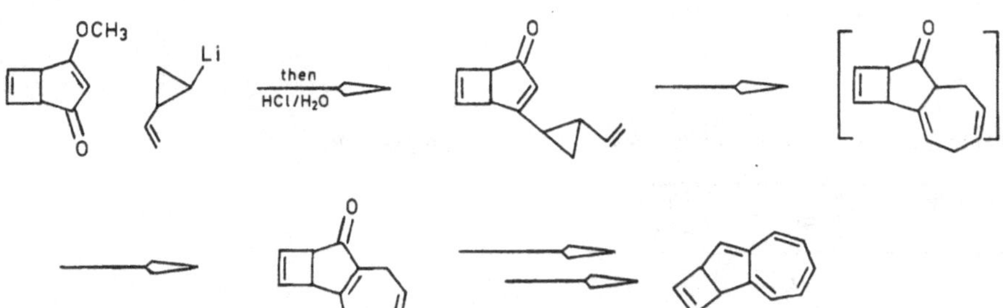

(21)

To direct a solvolytic ring opening, 2-methoxycyclopropyllithium (*6*) was developed as a chain extension conjunctive reagent. The failure of β-elimination to occur in *6* presumably derives from the high strain of cyclopropene and poor orbital overlap for elimination. The aldehyde adducts smoothly solvolyze to give β,γ-unsaturated aldehydes (Eq. 22) [23] which are best initially isolated as their hemithioacetals.

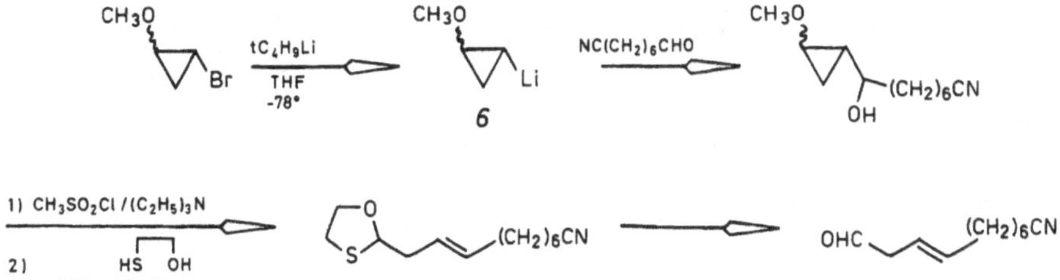

(22)

The cuprate derived from the cis isomer of this reagent undergoes conjugate addition with cyclohexenone with unusually high diastereoselectivity (5:1, Eq. 23)[24]. In this case, electrophilically initiated ring opening with mercuric acetate chemoselectively attacks the sterically least hindered cyclopropyl bond to give a branched product *7*. Reductive work-up produces *8* in which the stereochemistry of the

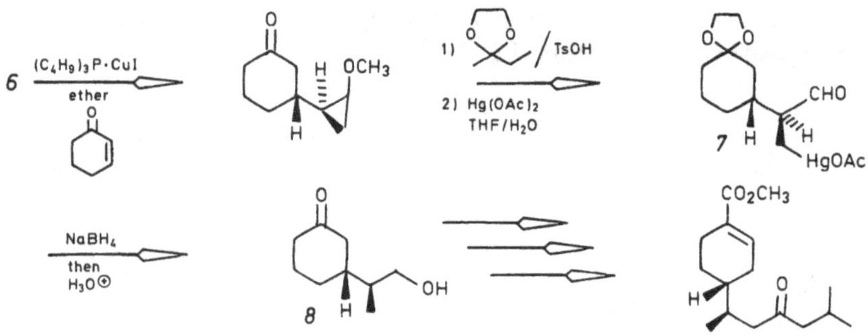

(23)

acyclic unit is controlled as a result of the initial conjugate addition.

It is obvious that the number of potentially useful conjunctive reagents has barely been touched. For the three discussed herein, the synthetic transformations developed so far are summarized in Table 1.

Table 1. Synthetic Constructions with Unstabilized Cyclopropyl Anions

Conjunctive reagent	Substrate	Final product[a]
	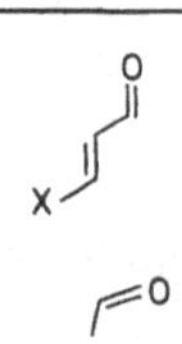	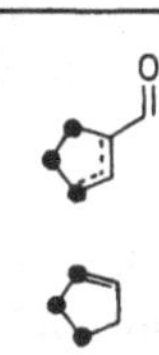
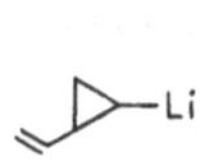	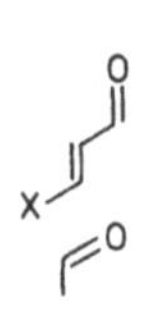	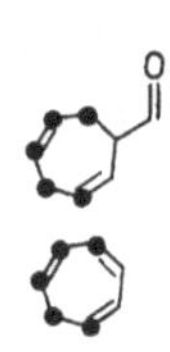
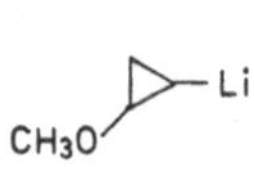	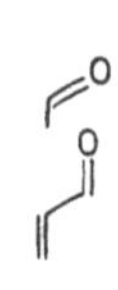	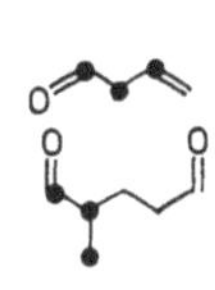

[a] The carbons dotted in the product derive from the cyclopropyl conjunctive reagent.

5 Sulfonium Ylides

Placing anion stabilizing groups on the cyclopropane greatly facilitates generation of cyclopropyl nucleophiles. The diphenylsulfonium ylide has proven to be an exceptionally versatile conjunctive reagent [25]. The sulfonium salt *9*, available from either 3-chloro-1-iodopropane (Eq. 24a) or 3-chloro-1-propanol (Eq. 24b), is a nicely crystalline stable salt that can be stored indefinitely [26,27]. While substituted

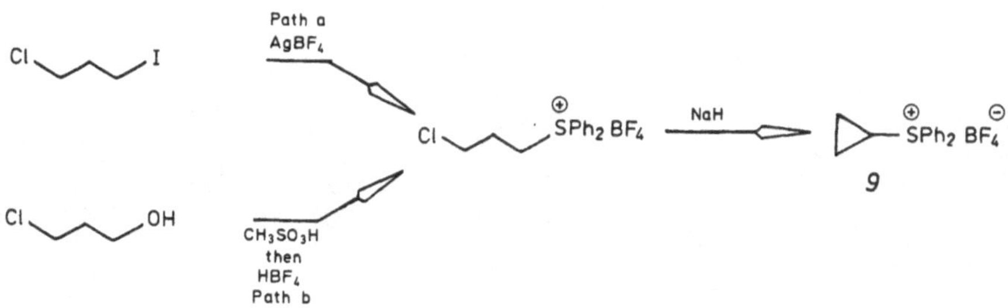

(24)

cyclopropyl salts can, in principle, be made by similar routes, only the methyl substituted salt *10* has been made to date [27]. The ylides are generated from

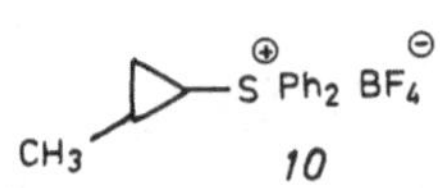

(24a)

the salts upon deprotonation under either irreversible (dimsylsodium in DME at −40 °C) or reversible (powdered potassium hydroxide in DMSO) conditions. Other ways have included potassium t-butoxide in THF and potassium hydroxide in acetonitrile. Synthetically, the reversible ylide generation conditions have proved to be most useful. The inversion barrier (Eq. 25) is higher than for cyclopropyl phenyl sulfone ($E_a = 71$ kJ/mole, 17 Kcal/mole)), but inversion, does occur in DMSO although not methanol.

(25)

Related to the sulfonium salts and their ylides are the oxosulfonium salts and their ylides. Cyclopropyl (dimethylamino)phenyloxosulfonium fluoroborate *11*, available by routes analogous to the preparation of *9*, suffers smooth deprotonation to the corresponding ylide upon treatment with base (Eq. 26) [28].

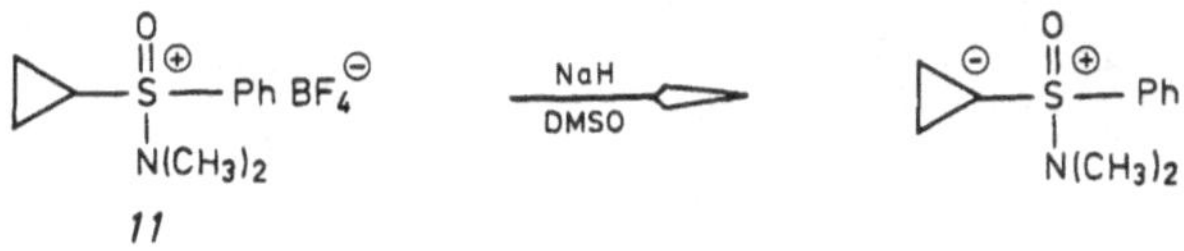

(26)

Table 2. Oxaspiropentanes and Cyclobutanones Using Sulfonium Cyclopropylides

Entry	Aldehyde or ketone	Salt, method[a]	Oxaspiropentane[a]		Rearrangement method[b]	Cyclobutanone		Yield[d]	Ref.
1	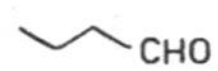	9, A	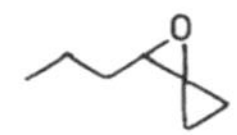	(100)					29)
2	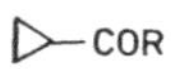	9, A				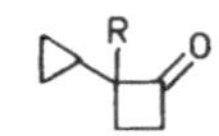	R = H	54	34)
							CH_3	86	35)
3	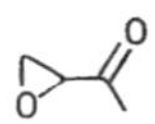	9, B	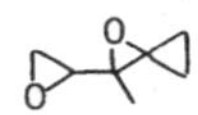	(71)	B	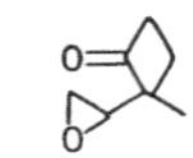		83	36)
4			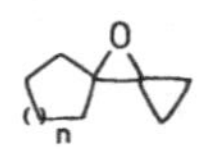			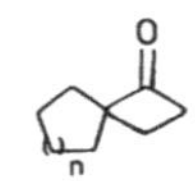			
a		9, A	n = 1	(94)	B	n = 1		94	35)
b		9, A	2	(97)	B	2		97	35)
c		9, A	3	(90)	B	3		90	35)
d		10, A	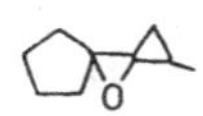		B	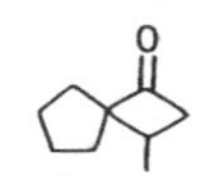		83	35)
5	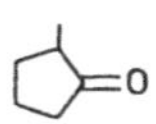	9, A	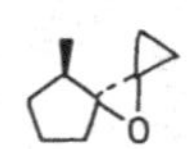		B	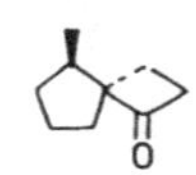		61	37)
6	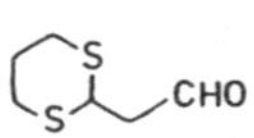	9, B	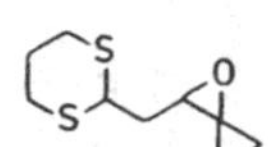		B	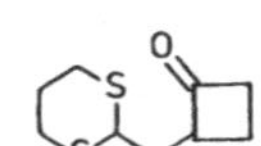		48	25c)

7		9, A	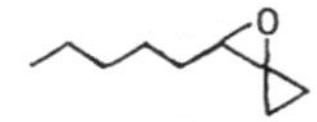	(92)	B	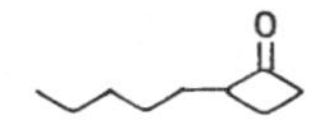		87	35)
8	PhCHO	9, A	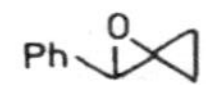	(87)	B	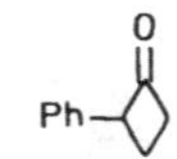		87	35)
9	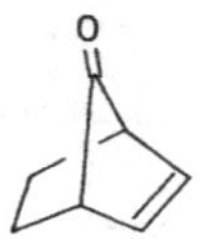	9				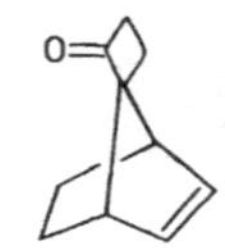 83	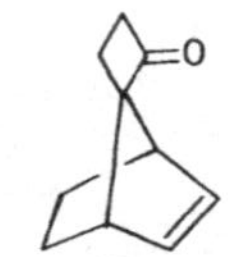 17		38)
10		9, A				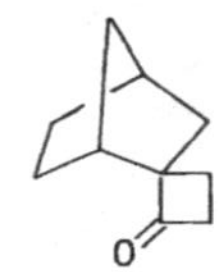		92	35)
					A	82	18		
					B	91	9		
11	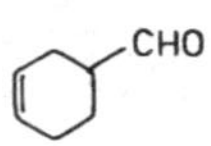	9, A	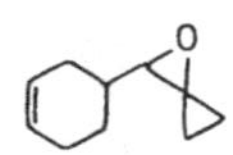	(80)	B	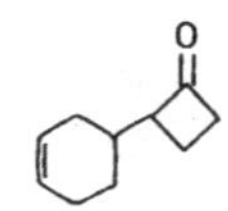		87	35)
12	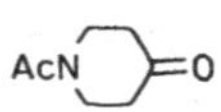	9, B	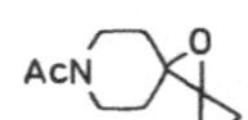		B	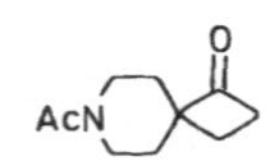		77	39, 40)
13	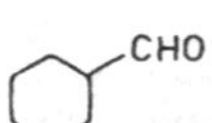	9, A	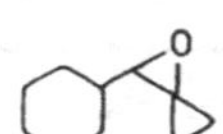	(80)					29)

Table 2. (continued)

Entry	Aldehyde or ketone	Salt, method[a]	Oxaspiropentane[a]	Rearrangement method[b]	Cyclobutanone	Yield[d]	Ref.
14	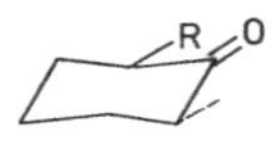	9, A	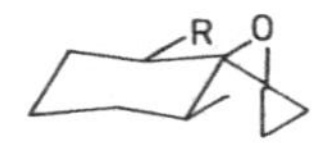	B	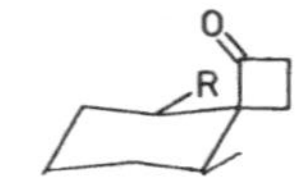 R = H CH_3	80 88	37)
15	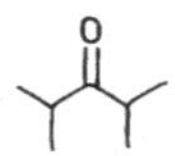	9, A	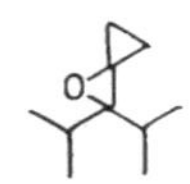 (59)	B	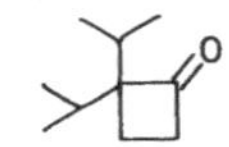	59	29)
16	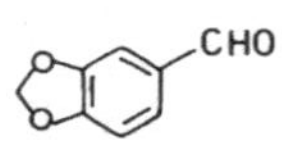	9, A			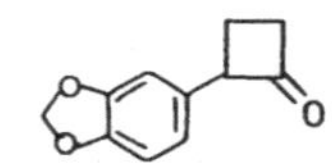	71	41)
17	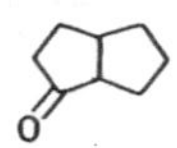	9, A	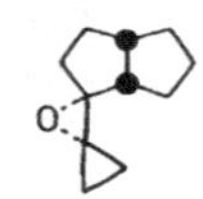 (76)	A	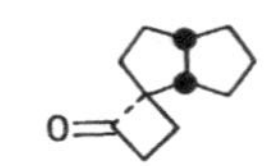	80	42)
18	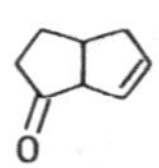	9, A	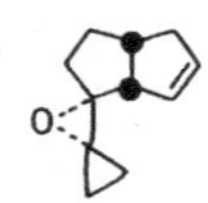 (>82)				43)
19	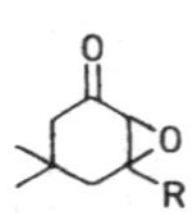		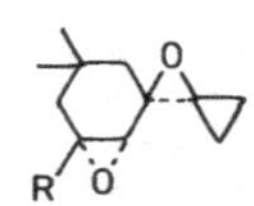		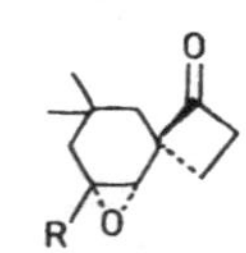		
a		9, A	R = H (96)	A		96	36)
b		9, A	R = CH_3 (88)	A		88	36)

20	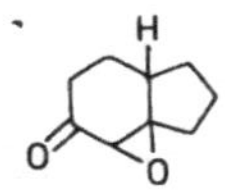	9, A	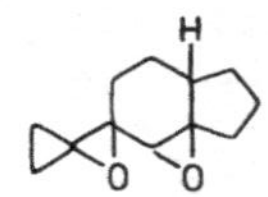 (85)	E	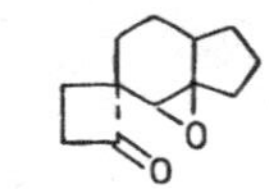 87 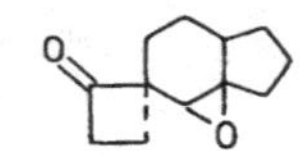 13	92	36)
21	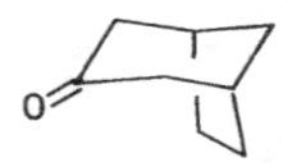	9, A	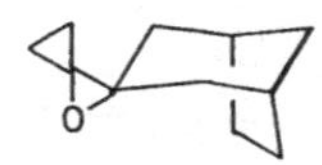	B	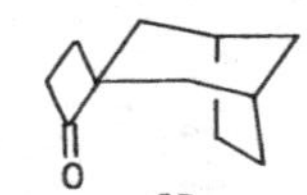 65 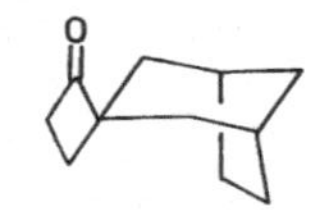 35	44	35)
22	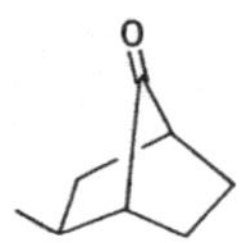	9, A			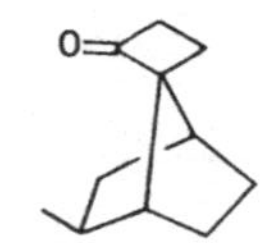	20	44)
23	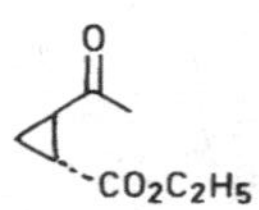	9	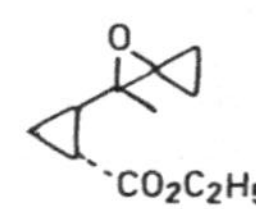		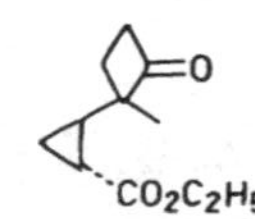	74	45,46)
24	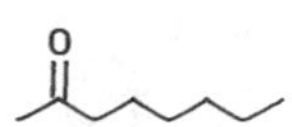	9, A	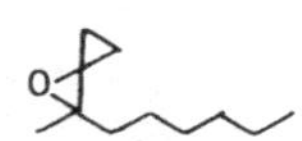 (92)	B	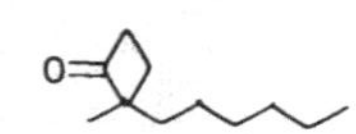	92	35)
25	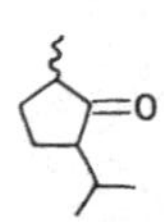	9, A	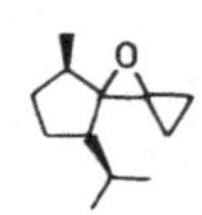	B	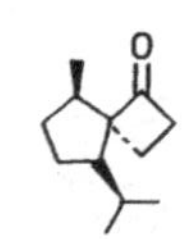	89	47)
26	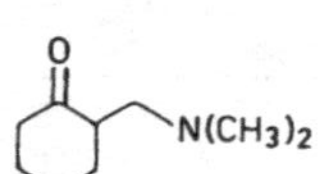	9, B	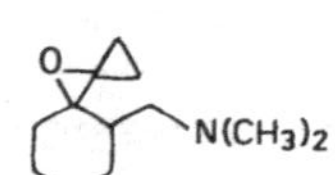 (44)	B			47)

Table 2. (continued)

Entry	Aldehyde or ketone	Salt, method[a]	Oxaspiropentane[a]	Rearrangement method[b]	Cyclobutanone		Yield[d]	Ref.
27	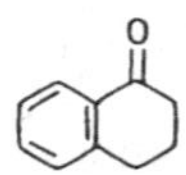	9, A			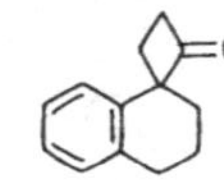		72	35)
28	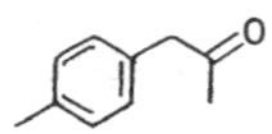	9, A	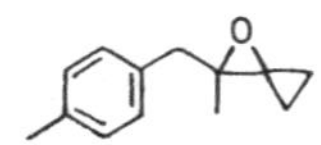 (85)	,	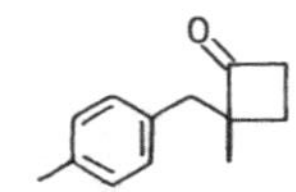			49)
29	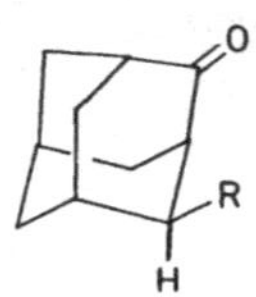	9, A	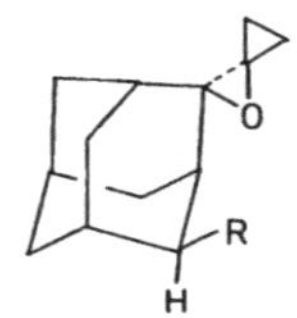	A	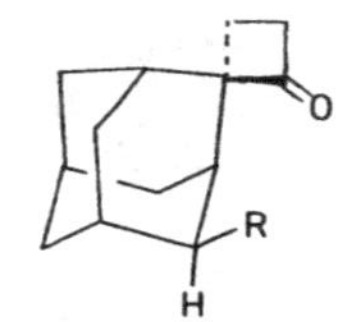	R = H	92	50)
						CH_3	53	51)
30		9, A	 (80)				80	35)
a				A	82	18		
				B	91	9		
				C	> 99	< 1		
b		10			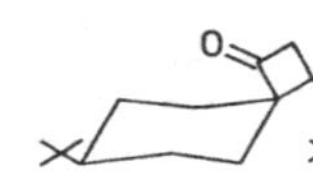 63	 16	95	35)
					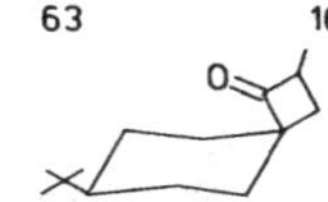 21			

31	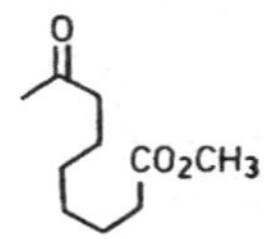	$\underline{9}$, A	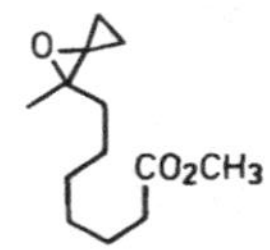	B	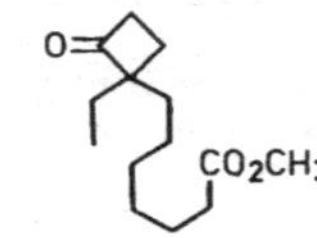	57	49)
32	Ph_2CO	$\underline{9}$, A			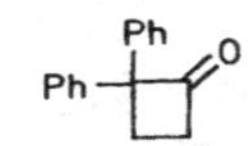	91	35)
33	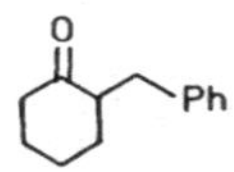	$\underline{9}$, A	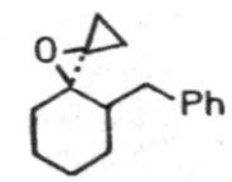	B	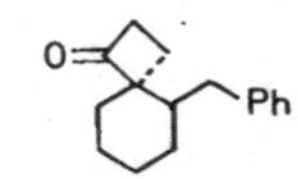	74	39)
34	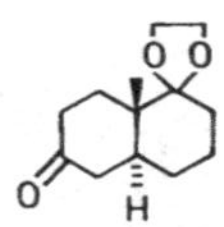	$\underline{9}$, A	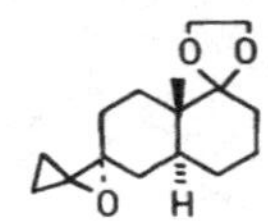	B	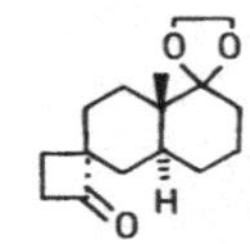	30	39)
35	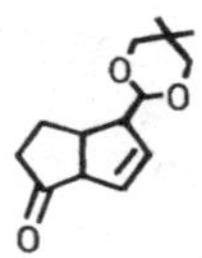	$\underline{9}$, A	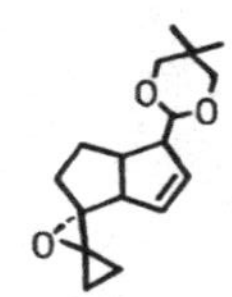 (> 71)				42)
36	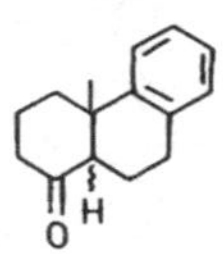	$\underline{9}$, A	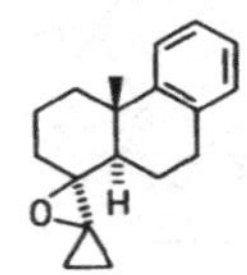	B	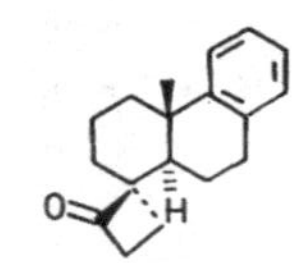	80-99	35)

Table 2. (continued)

Entry	Aldehyde or ketone	Salt, method[a]	Oxaspiropentane[a]	Rearrangement method[b]	Cyclobutanone	Yield[d]	Ref.
37	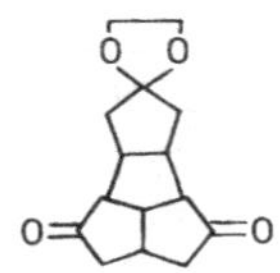	9, C	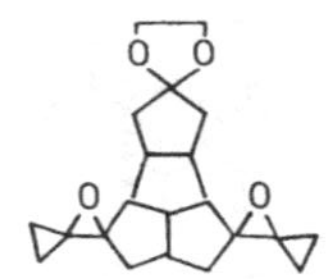	D	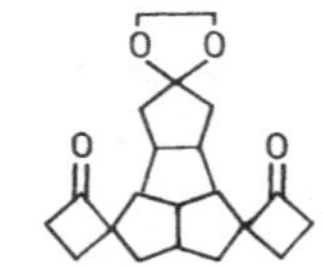	70	52)
38	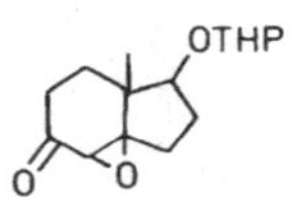	10, A	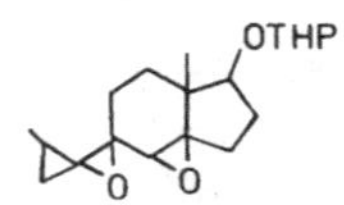	E	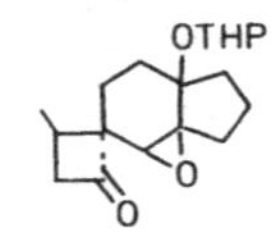	79	36)
39	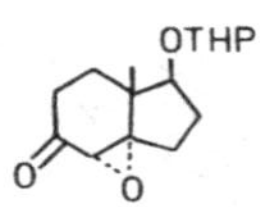	10, A	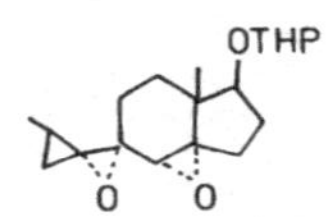	E	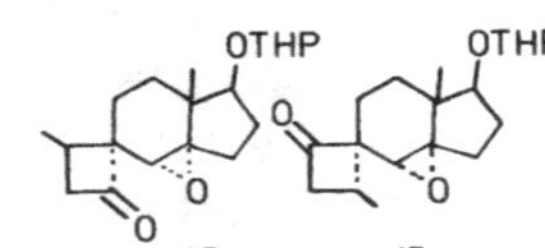 85 15	80	36)
40	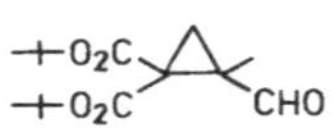	9, A	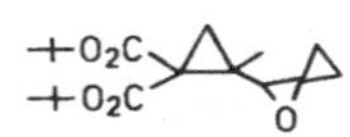	B	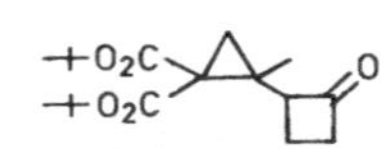	25	46)
41	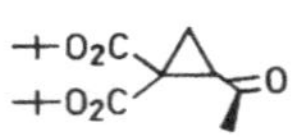	9, D	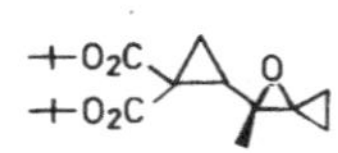	B	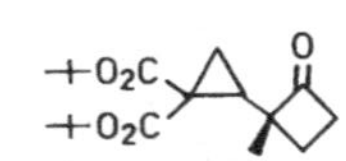	82	46)
42	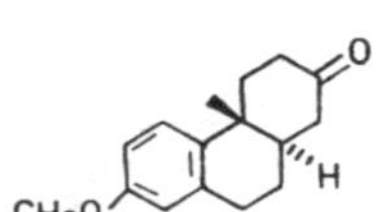	9, A	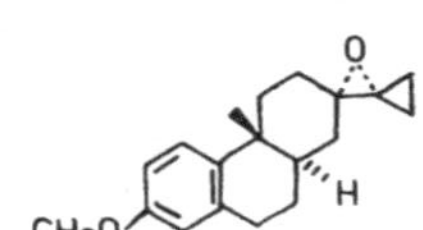	B	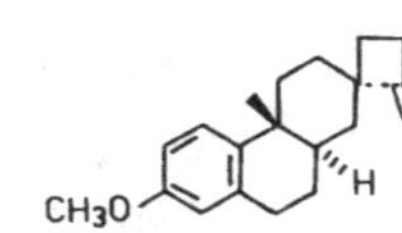	70-5	39)

43	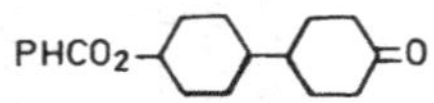	9, A	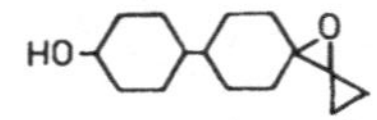	B	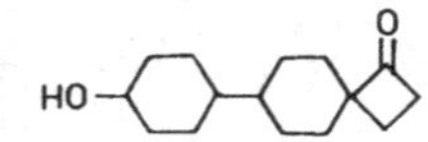	50	39)
44	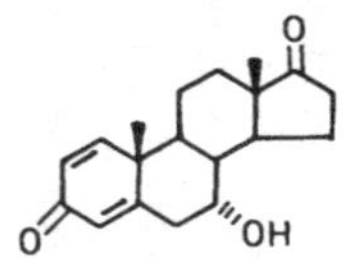	9			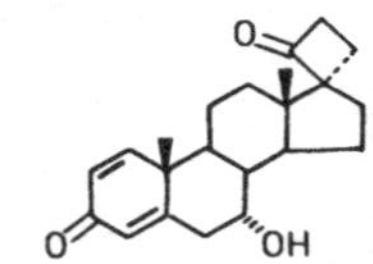	U	53)
45	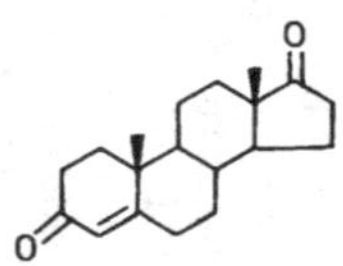	9			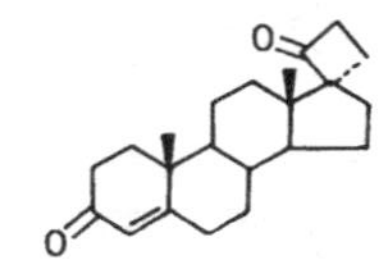	U	54)
46	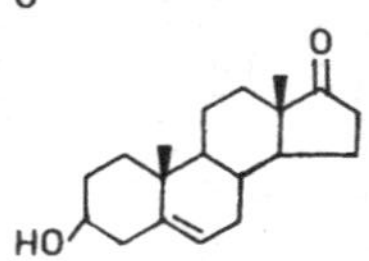	9, A	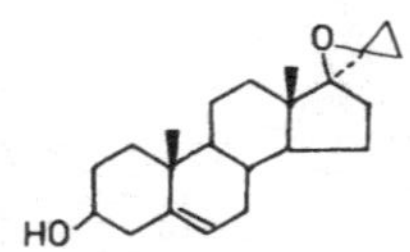	B	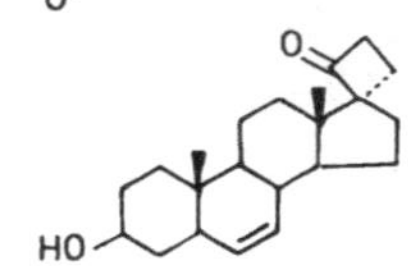	67	37)
47	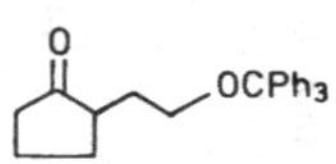	9, A	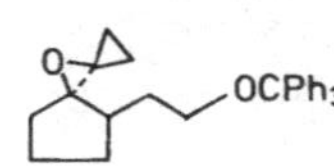	B	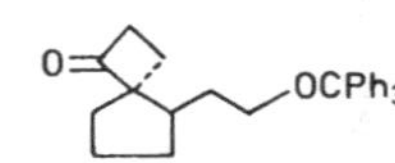	81	55)
48	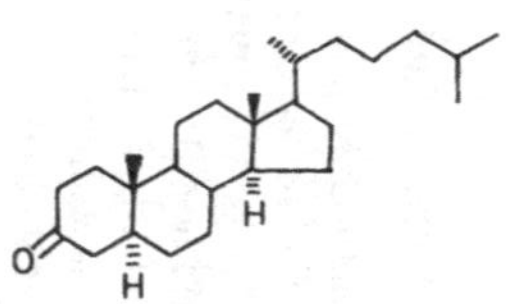	9, A	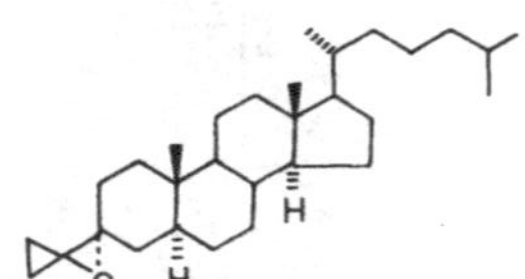	B	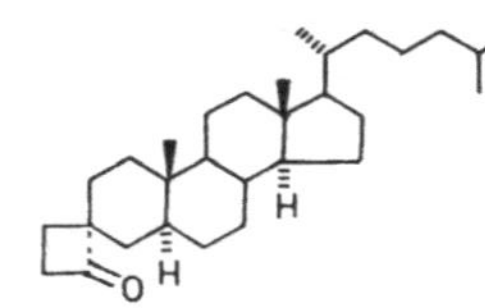	51	

The most remarkable feature of the chemistry of these ylides is their efficient participation in typical sulfur ylide chemistry, i.e. their ability to form epoxides with carbonyl partners (Eq. 27a)[29] and spiropentanes with enones (Eq. 27b)[30].

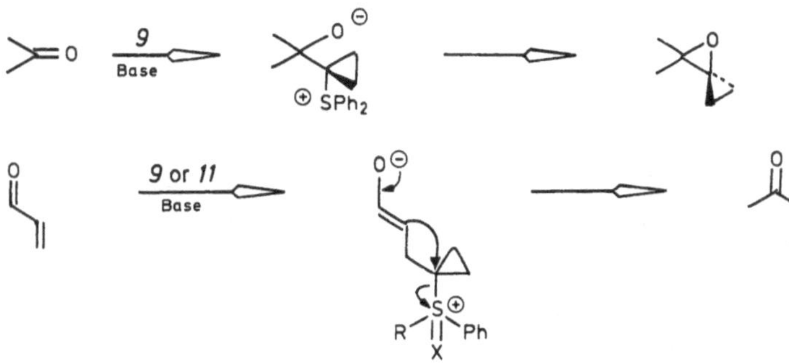

(27a)

(27b)

These reactions, which involve a normal S_N2 displacement as depicted, represent the first authenticated examples of displacement with inversion of configuration at a cyclopropyl carbon[27,31]. The facility of these reactions (proceeding at temperatures below 0 °C) seems even more remarkable considering the distortion that must accompany the flattening of a cyclopropyl carbon.

A related series of reactions derives from dibromocarbene adducts of olefins as in Eq. 28[32,33]. Metal-halogen exchange at very low temperature produces the

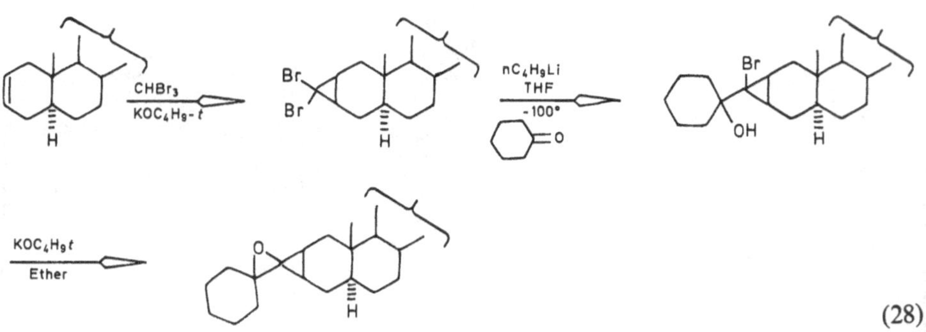

(28)

Footnotes to Table 2.

[a] Method A: powdered KOH in DMSO; Method B: dimsylsodium in DME; Method C: powdered KOH in acetonitrile; Method D: potassium t-butoxide in DMSO at room temperature

[b] Method A: aqueous fluoroboric acid and pentane or carbon tetrachloride at room temperature; Method B: lithium fluoroborate or perchlorate in PhH at room temperature to reflux; Method C: $Eu(fod)_3$ in deuteriochloroform at 37 °C; Method D: TsOH in acetonitrile; Method E: oxalic acid in acetonitrile

[c] Yields in parenthesis are for isolated pure oxaspiropentanes. In cases where a yield is not given, the crude oxaspiropentane was directly employed in subsequent reactions. In cases where no structure is given, the oxaspiropentane rearranges under conditions of its formation to give the cyclobutanone directly

[d] Yields are for isolated cyclobutanone based upon starting carbonyl partner.

α-bromocyclopropyllithium reagent which undergoes normal carbonyl additions. Analogous to Eq. 27a, base treatment completes formation of an oxaspiropentane also by a supposedly unfavorable displacement at a cyclopropyl carbon. In these cases, the oxaspiropentanes are normally immediately rearranged to cyclobutanones (vide infra).

The direct high yield synthesis of oxaspiropentanes from almost any type of aldehyde or ketone represents a particularly useful transformation because of the high reactivity of such compounds. This approach proves to be exceptionally simple. The DMSO reaction mixture can be directly extracted with pentane or hexane, the hydrocarbon solvent removed and the product isolated by distillation or crystallization. Since diphenyl sulfide is the only by-product extracted with the oxaspiropentane, the mixture can normally be used for most further synthetic transformations. Table 2 summarizes some of the oxaspiropentanes prepared by this method.

Several aspects are particularly noteworthy. Good chemoselectivity is noted in the compatibility with epoxides, esters, olefins, and alcohols. Entries 44 and 45 demonstrate the chemoselectivity between an unsaturated and saturated ketone.

The diastereoselectivity is striking. Even when steric factors are not overwhelming (eq. Table 2, entries 5, 9, 22, 29, 30, and 33) only a single oxaspiropentane was detected. A particularly useful aspect of this reaction deals with carbonyl partners that are easily epimerized at the α-carbon. It appears that epimerization is faster than carbonyl addition. However, since one of the two epimers reacts faster than the other, only a single diastereomeric oxaspiropentane still results. For example, 2-isopropyl-5-methylcyclopentanone exists as an *E,Z*-mixture (see Eq. 29) [47]. For steric reasons, the *Z* isomer reacts faster than the *E* isomer which leads to *12* as the

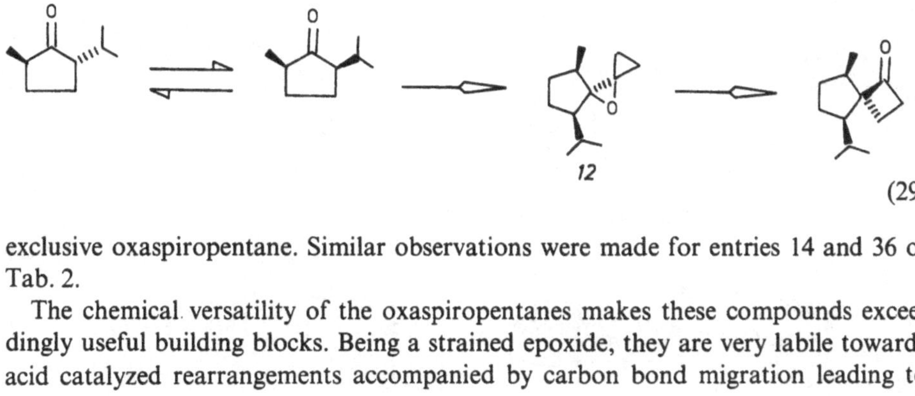

12

(29)

exclusive oxaspiropentane. Similar observations were made for entries 14 and 36 of Tab. 2.

The chemical versatility of the oxaspiropentanes makes these compounds exceedingly useful building blocks. Being a strained epoxide, they are very labile towards acid catalyzed rearrangements accompanied by carbon bond migration leading to

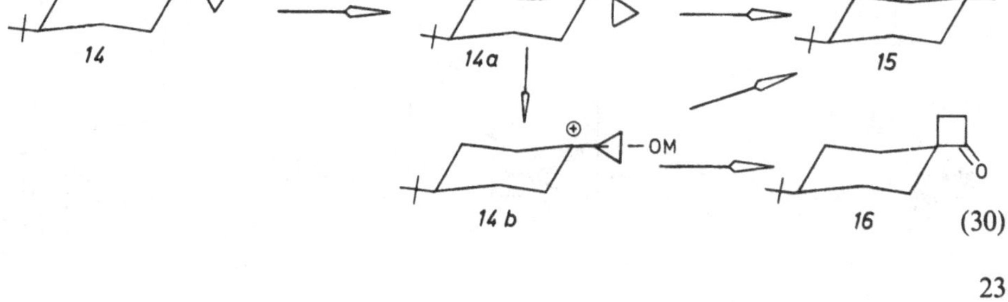

14 *14a* *15* *14b* *16*

(30)

cyclobutanones [56, 57]. The diastereoselectivity depends upon the acid employed for the rearrangement. For example, protonic acids rearranged *14* (Eq. 30) to give an 82:18 ratio of *15*:*16* which increased to 91:9 by using a Li (+1) salt and to >99:1 by using an Eu (+3) salt. This stereochemical event reflects the rate of rearrangement of *14a* to cyclobutanone *15* compared to the rate of bond rotation to *14b* which can produce either *15* or *16*. Relaxation of *14a* to the presumed thermodynamically more stable *14b* should be fast. Apparently, by making the oxygen strongly electron releasing, rearrangement can dominate. The formation of *15* from *14* represents a migration of the cyclopropyl bond with inversion of configuration at the migration terminus — a result also consistent with this process being concerted.

The facility of the rearrangement to cyclobutanones is reflected in the high chemoselectivity. The cases of oxaspiropentanes from epoxyketones offer a particularly difficult challenge. Nevertheless, no problems resulted (Table 2, entries 19, 20, 38 and 39). Oxaspiropentanes which form particularly stabilized carbonium ions frequently rearrange to cyclobutanones during their formation. For example, cyclopropylmethyl ketone and benzophenone led only to cyclobutanones in their condensations with *9*. In one case, further reaction of the ylide with the rearranged cyclobutanone was noted (Eq. 31) [58].

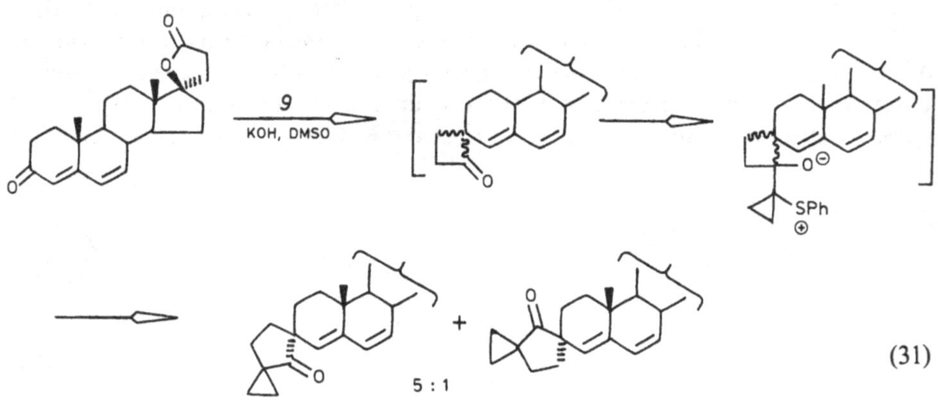

(31)

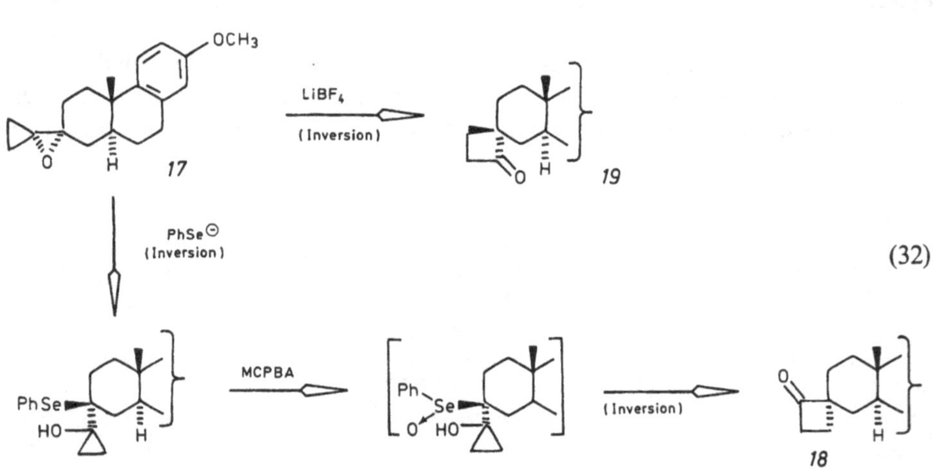

(32)

In order to accomplish the alternative stereochemical result, a process which invokes two inversions from *14* would be required. Equation 32 illustrates such a process using the highly nucleophilic selenide anion for the first inversion [37]. Without isolation, the hydroxyselenide is exposed to an oxidant. The derived selenoxide cannot be isolated since ionization, facilitated by the neighboring group effect of the hydroxycyclopropyl ring, leads to ring enlargement also predominantly as an inversion process. Table 3 summarizes the few examples. Thus, a single oxaspiropentane *17* creates either epimeric cyclobutanone *18* or *19* simply by choosing the proper rearrangement conditions.

Substituted cyclopropyl ylides also participate in oxaspiropentane formation (Table 2, entries 4d, 30b, 38, and 39). Of the two cyclopropyl carbons that can move in the rearrangement to cyclobutanones, the carbon that best stabilizes a

Table 3. Stereoreversed Cyclobutanone Formation from Oxaspiropentanes

Entry	Oxaspiropentane	Cyclobutanone		Yield	Ref.
	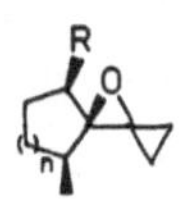	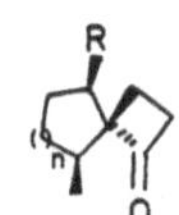	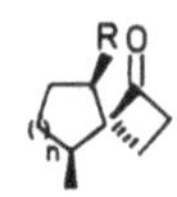		
1	n = 1, R = H	96	4	48	37)
2	n = 2, R = H	87	13	85	37)
3	n = 2, R = CH_3	40	60	85	37)
4	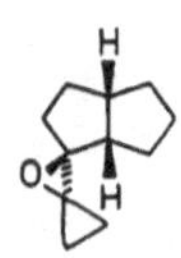	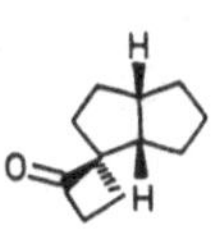		70	42)
5	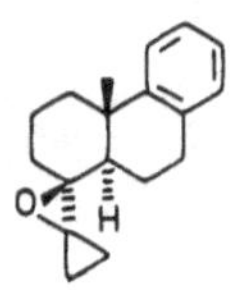	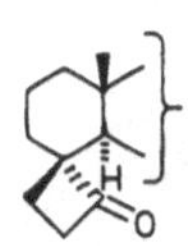 64	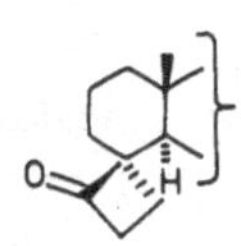 36	68	37)
6	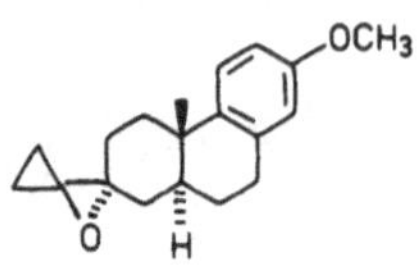	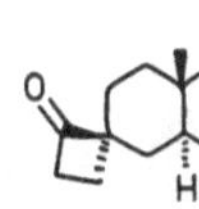 93	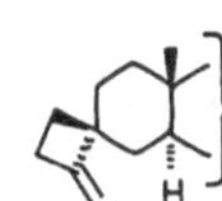 7	65	37)
7	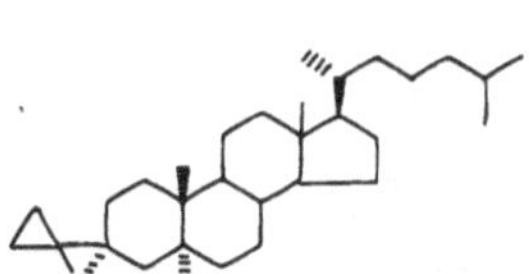	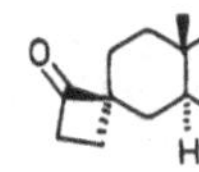 > 98	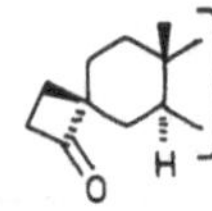 < 2	62	37)

Table 4. Preparation of Vinylcyclopropanol Silyl Ethers from Oxaspiropentanes

Entry	Oxaspiropentane	Base	Vinylcyclopropanol silyl ether[a]	Yield
	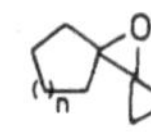	$LiN(C_2H_5)_2$	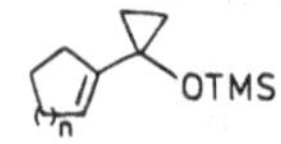	
1	n = 1		n = 1	94
2	n = 2		n = 2	I[b]
3	n = 3		n = 3	91
4	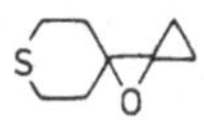	$LiN(C_2H_5)_2$	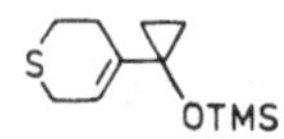	60
	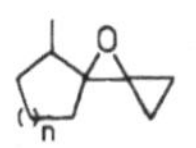	$LiN(C_2H_5)_2$	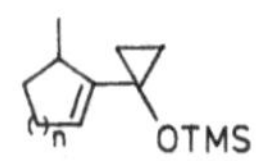	
5	n = 1		n = 1	I[b]
6	n = 3		n = 3	I[b]
	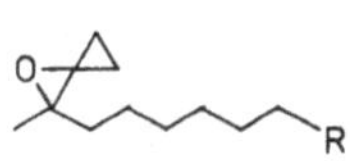		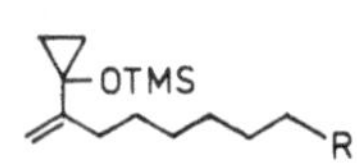	
7	R = H	$LiN(C_2H_5)_2$	R = H	I[b]
8	$R = CO_2C(CH_3)_3$	$LiN(iC_3H_7)_2$	$R = CO_2C(CH_3)_3$	Q[c]
	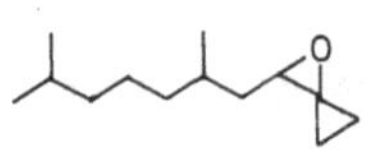	$LiN(C_2H_5)_2$	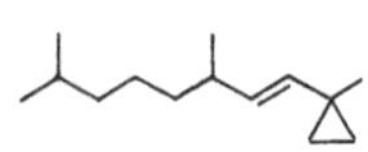	I[b]
10	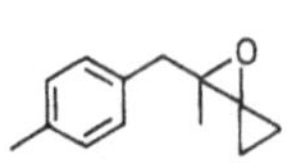		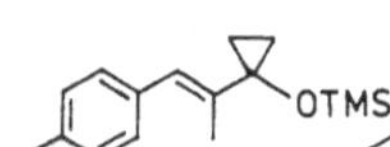 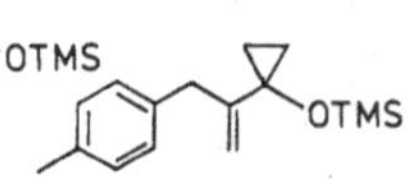	
		a) $LiN(iC_3H_7)_2$	99 1	I[b]
		b) $LiN(C_2H_5)_2$	81 19	
		c) 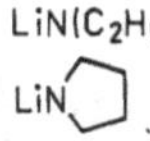	— >98	I[b]
	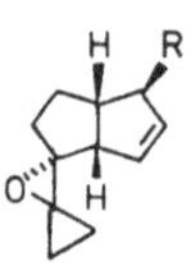	$LiN(C_2H_5)_2$	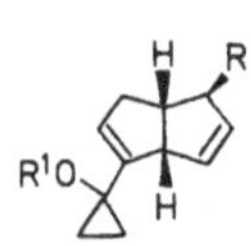	
11	R = H		a) R = H, R^1 = TMS	70%
			b) R = R^1 = H	88%
12	R = 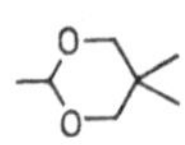		R = 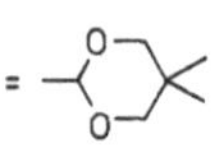 R^1 = H	>77%[d]
13	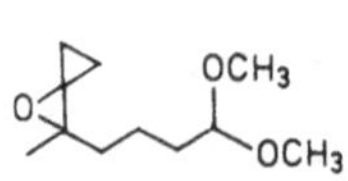	$LiN(C_2H_5)_2$	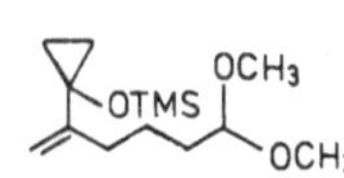	>77%[d]

developing positive charge, i.e. the substituted cyclopropyl carbon, preferentially migrates.

The sequence of ketone to cyclobutanone represents a replacement of each C—O

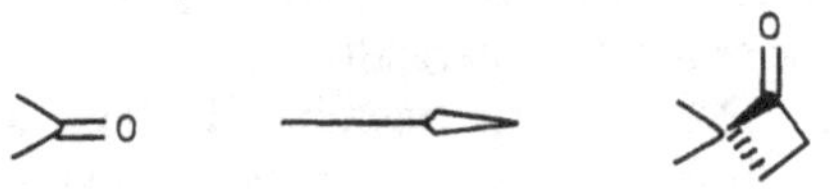

(32a)

bond of the carbonyl group by a C—C bond – i.e. a geminal alkylation. Furthermore, high diastereoselectivity accompanies this reaction. A related sequence utilizes the cyclopropyl phosphorus ylide [59] to form alkylidenecyclopropanes as in Eq. 33 [60]. Epoxidation and rearrangement of the derived oxaspiropentane [61]

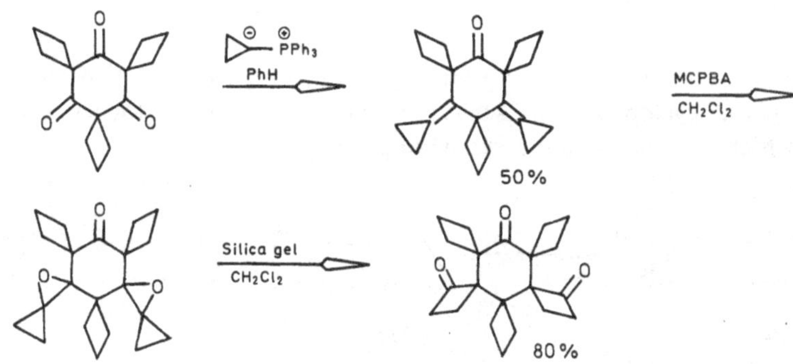

(33)

completes this sequence. The low yields of the phosphorus ylide reaction and the lack of stereocontrol in the epoxidation limit the usefulness of this multi-step sequence.

A second major reaction of oxaspiropentanes as reactive epoxides is their elimination to form vinylcyclopropanols [29, 49, 62]. A rapid elimination to vinylcyclopropanols occurs when the oxaspiropentanes are exposed to lithium dialkylamides in hexane or pentane as exemplified in Eq. 34 and Table 4.

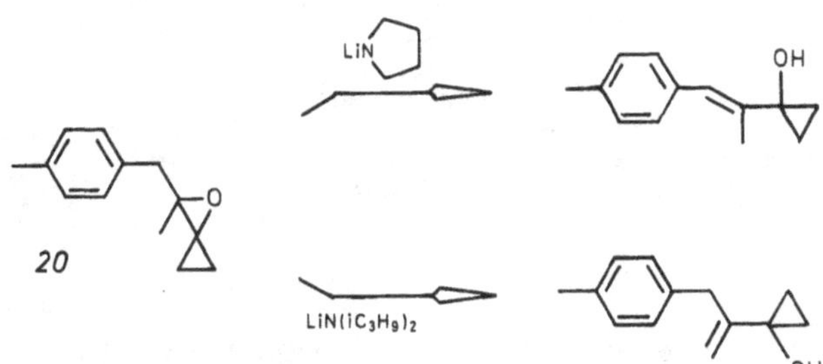

20

(34)

Footnote to Table 4.

[a] The pentane or hexane solution of the lithium alkoxide is diluted with DME before addition of TMS-Cl for silylation.

[b] In most cases, the yields are indeterminate (I) since crude oxaspiropentanes were employed directly and the resultant vinylcyclopropanol silyl ethers used crude in subsequent reactions.

[c] Quantitative yield.

[d] Yield quoted is overall from starting ketone.

The reaction presumably involves a cis, syn elimination. As Eq. 34 illustrates, regioselectivity can be controlled by choice of base [49]. The higher kinetic acidity of the benzylic position of *20* determines the regioselectivity with a non-hindered base; whereas, steric hindrance directs the base to the methyl group.

The reaction fails if the proton to be removed is sterically hindered — either tertiary as in *21* or neopentyl-like as in *22* [63]. In six membered rings, the cis, syn hydrogen must be axial for elimination. In the parent cyclohexyl system *23*, a mixture of *24* and *25* results [29]; whereas, in a conformationally rigid cyclohexane

(34a)

ring corresponding to *23a*, no vinylcyclopropanol forms [65]. The formation of the desired vinylcyclopropanol *25* may derive from reaction via the less stable conformer *23b* which does posses a cis, syn axial proton.

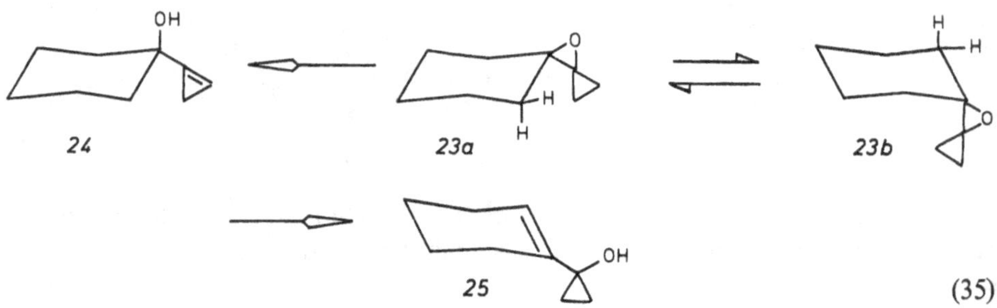

(35)

An alternative procedure to effect elimination resolves this problem. Opening the oxaspiropentane *26* with selenide anion in a non-protic solvent effects a direct elimination via a merged substitution — elimination mechanism to give the vinyl-

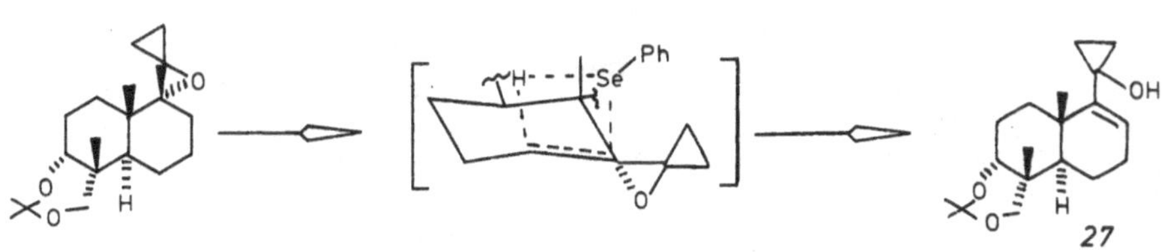

(36)

cyclopropanol *27* in 56% yield [65]. Thus, by switching from a cis, syn to a trans, anti mechanism, the reaction allows for effective overlap of the cleaving axial proton with the breaking axial C—O bond. While this approach has not been explored generally, its effectiveness in *26* where no elimination occurs whatsoever by the standard base method suggests it may be a powerful technique. The importance of solvent becomes obvious by contrasting the formation of *27* in this case compared to smooth hydroxyselenide formation in ethanol — a key intermediate in the stereoreversed cyclobutanone spiroannulation (vide supra).

6 Metalated Thioether, Selenoether, and Ether

A companion reagent to the ylide is the corresponding metalated sulfide *30* which arises by the lithiation of *29* with n-butyllithium [66]. The latter forms by base closure of *28*. Since closure of *28b* involves use of n-butyllithium, the cyclopropyl sulfide *29* simply becomes an intermediate which is metalated in situ to give *30* directly [67]. A non-metalation sequence involves *32* which undergoes reductive

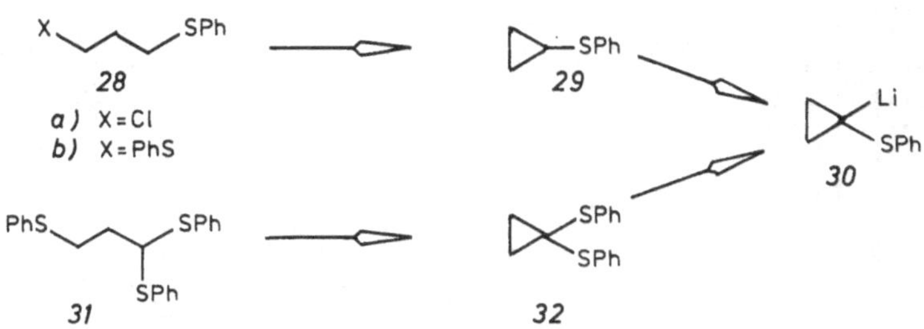

cleavage with lithium arylides, preferably that derived from 1-dimethylaminonaphthalene, to produce *30* [68]. The availability of *32* from the thiophenol adduct of acrolein, i.e. *31* [69], by reaction of n-butyllithium makes it an attractive alternative.

Similar methods also generate the selenium and oxygen analogues *33* [70] and *34* [71], respectively, as outlined in Eq. 37 and 38. These organolithium species add to carbonyl groups of non-conjugated and conjugated aldehydes and ketones, frequently

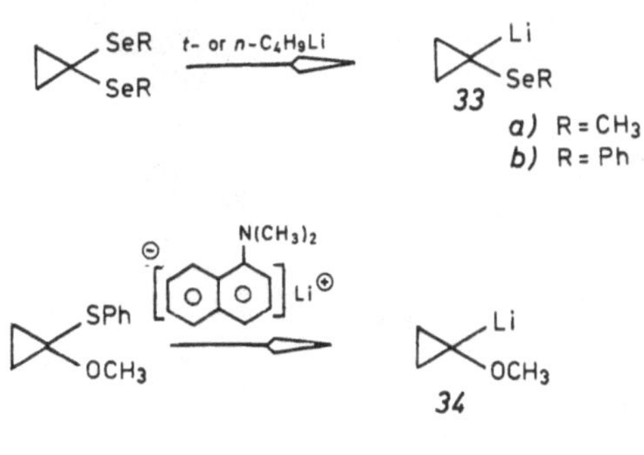

(37)

(38)

with good diastereoselectivity as summarized in Table 5. In contradistinction to an unsupported claim [85], in cases where comparisons between the sulfur and the selenium analogues can be made (cf. entries B-15 to B-20, B-29 and B-30), the sulfur analogue gives better yields of adducts. Stereochemistry derives from attack on the least hindered face [23,74,86]. Conjugate addition does not compete with direct carbonyl addition.

Table 5. Preparation of Cyclopropylcarbinols and Cyclobutanones Using Lithiated Heteratom Substituted Cyclopropanes

Entry	Aldehyde or ketone	Reagent	Adduct (% Yield)	Rearrangement method	Cyclobutanone (% Yield)	Ref.
			A. Non-conjugated aldehydes			
1	CH_2O	*30*	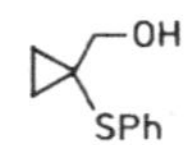 (80%)	A	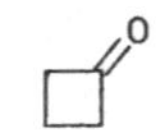 (70%)	72)
2	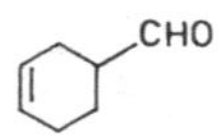	*30*	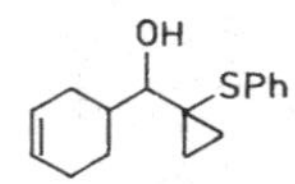 (93%)			73)
3	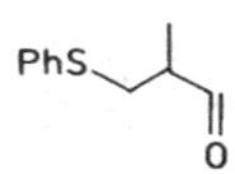	*30*	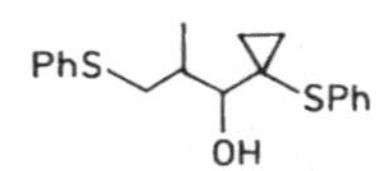 (93%)	B	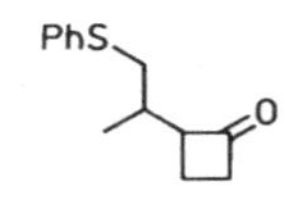 (90%)	73, 74)
4	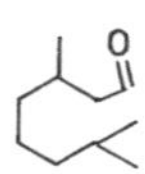	*30*	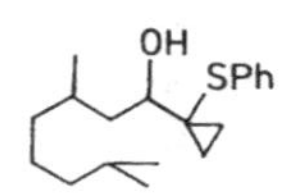 (91%)			73)
5	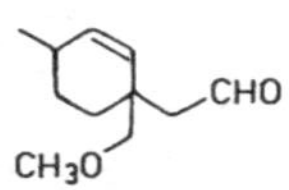	*30*	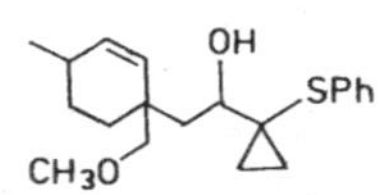	B	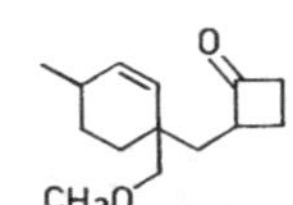 (62%)	75)
	RCHO		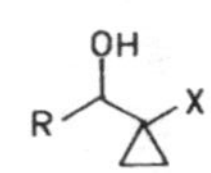		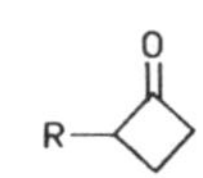	
6		*33*	X = PhSe R = C_6H_{13} (72)			70)
7			X = CH_3Se R CH_3 (69)			70)
8			C_6H_{13} (75)			70)
9			$C_{10}H_{21}$ (40, 75)	B	(73)	70)
10		*30*	X = PhS R = nC_5H_{11} (U)[6]			76)

Entry	Aldehyde or ketone	Reagent	Adduct (% Yield)	Rearrangement method	Cyclobutanone (% Yield)	Ref.
B. Saturated ketones						
1		*30*	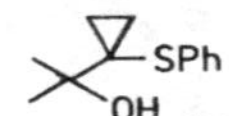 (56)			73)
2	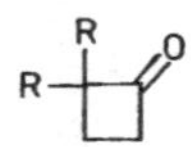	*30*	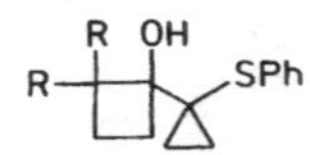	C	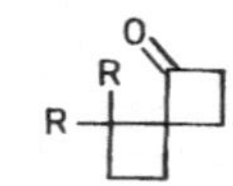	
			R = H (93)		(96)	73,74)
			R = CH_3 (97)			73)
3	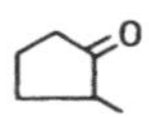	*30*	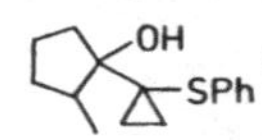 (41)			73)
	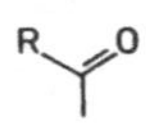		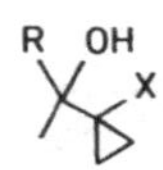		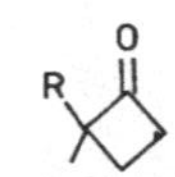	
4	R = nC_3H_7	*33*	X = CH_3Se (64)			70)
5	R = tC_4H_9	*30*	X = PhS (92)	B	(96)	73,74)
6	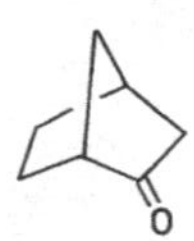	*30*	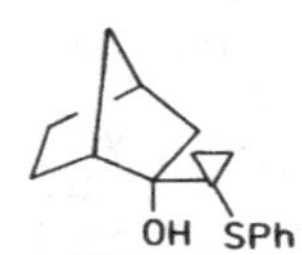 (93)	D	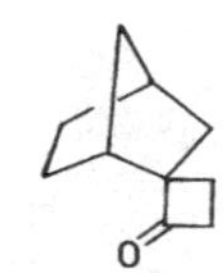 $(89:11)^d$ (63)	73,74)
7	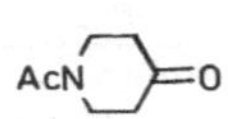	*30*	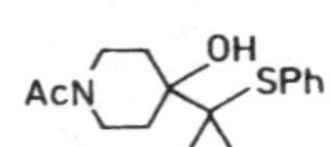 (25)			73)

Table 5. (continued)

Entry	Aldehyde or ketone		Reagent	Adduct (% Yield)		Rearrangement method	Cyclobutanone (% Yield)	Ref.
	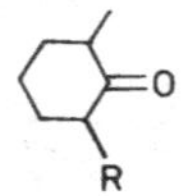		30	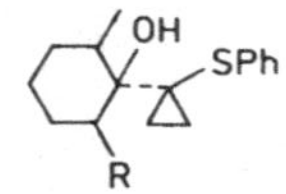				
8				R = H	(93)			73)
9				R = CH_3	(92)			73)
	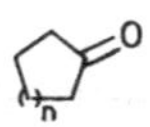		30	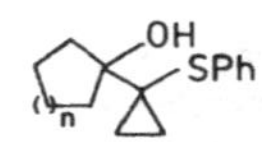			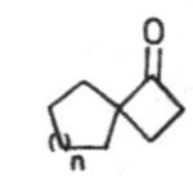	
10	n = 1			$(U)^b$				76)
11	n = 2			$(U)^b$				76)
12	n = 3			(51)		D	(84)	73,74,76)
13	n = 4			(67)		D	(100)	73,74)
14	n = 8			(94)				73 74)
	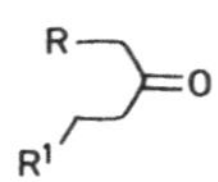			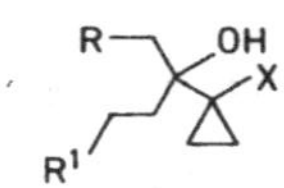			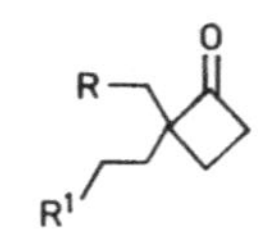	
15	R = CH_3	R^1 = H	33	X = CH_3Se	(55)			70)
16	C_2H_5	H	30	PhS	(87)			73)
17	H	nC_3H_7	30	PhS	$(U)^b$			76)
18	H	nC_4H_9	30	PhS	(87)			73)
19	H	$(CH_2)_6CO_2C_2H_5$	30	PhS	(87)			73)
20	H	nC_7H_{15}	33	CH_3Se	(70)	B	(73)	77)
21	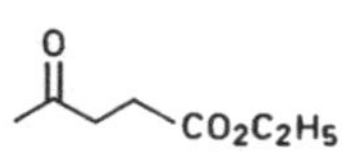		30	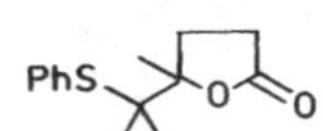	(46)			73)

	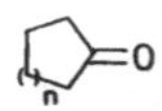		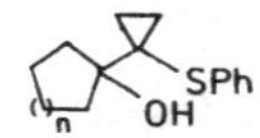		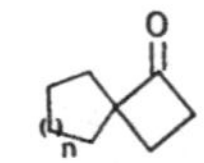	
22	n = 1		(U)[b]			76)
23	n = 2		(52)			68,73,76)
24	n = 3		(51)	D	(84)	68,73,76)
25	n = 4		(67)	D	(100)	73, 74)
26	n = 8		(94)			73)
27	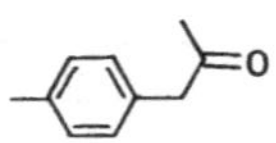	*30*	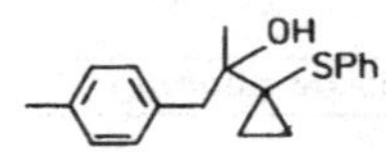 (93)			73)
28	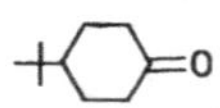	*30*	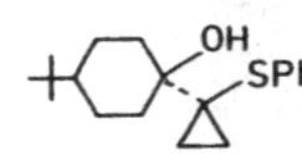 (95)		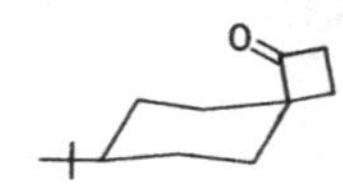 	
				C	7 — 93 (90)	73,74)
				E	45 — 55 (U)[b]	
	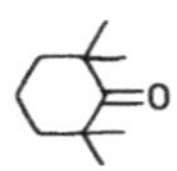		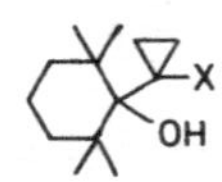			
29		*30*	X = PhS (98)			73)
30		*33*	X = CH_3Se (64)			77)
31	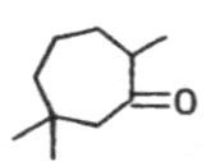	*30*	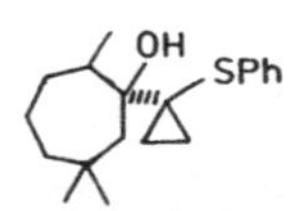 (62)			73)
32	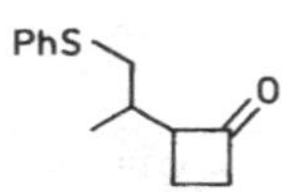	*30*	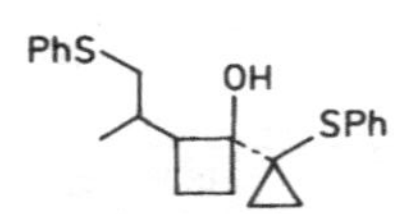 (89)	B	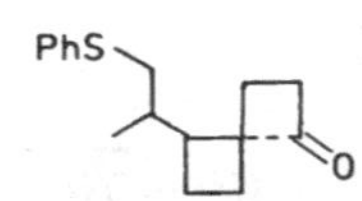 (92)	74)

Table 5. (continued)

Entry	Aldehyde or ketone	Reagent	Adduct (% Yield)	Rearrangement method	Cyclobutanone (% Yield)	Ref.
33	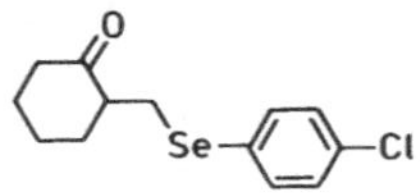	*30*	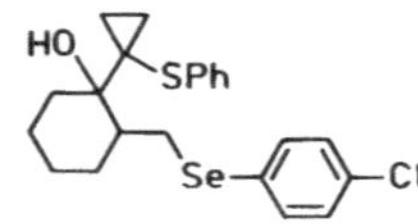		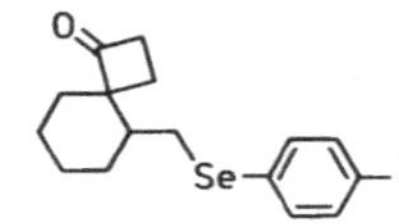 (76)[c]	78)
34	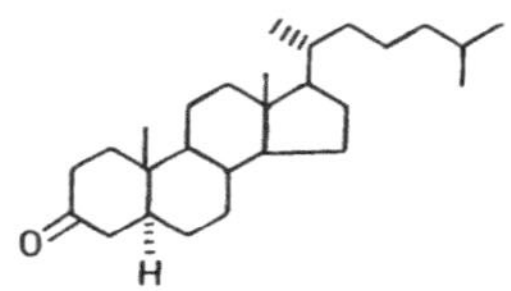	*30*	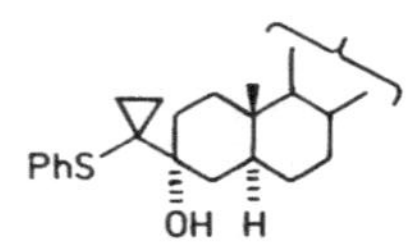 (93)			73)
35	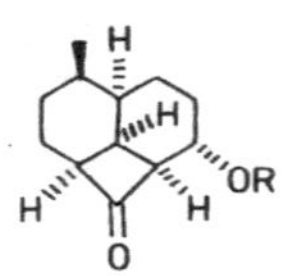 R = Si(CH$_3$)$_2$(tC$_4$H$_9$)	*30*	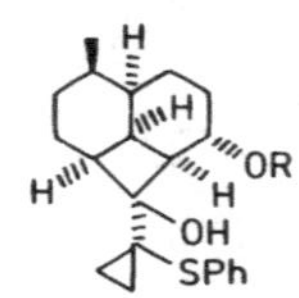	C	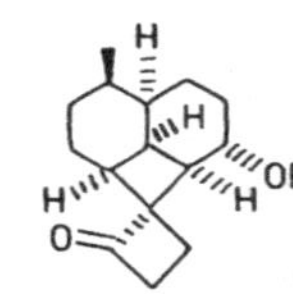 R = COPh (52)[c]	79)
	C. Conjugated carbonyl partners					
1	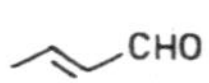	*30*	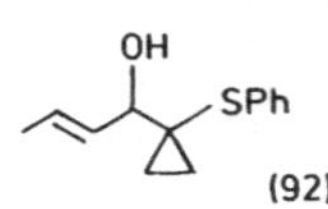 (92)	C E	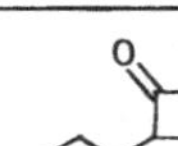 (95) (84)	73, 74)
2	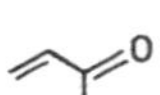	*30*	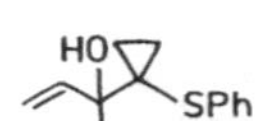 (83)	D	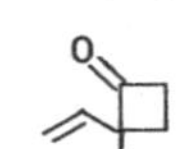 (85)	73, 74)
3	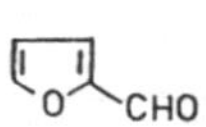	*34*	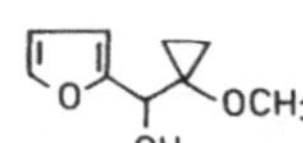	D	 (54)	80)

	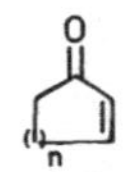		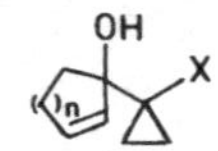		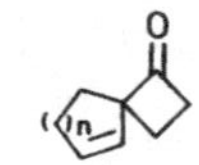	
4	n = 1	*30*	X = PhS (U)[b]			81,82)
5	n = 2	*30*	X = PhS (85)	D	(91)	73,74)
6	n = 3	*30*	X = PhS (U)[b]			82)
7	n = 3	*34*	X = CH_3O	D	(55)[c]	83)
8	PhCHO	*30*	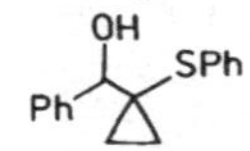 (71)			68)
	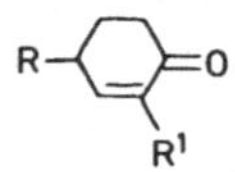		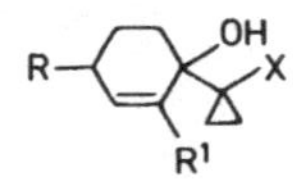		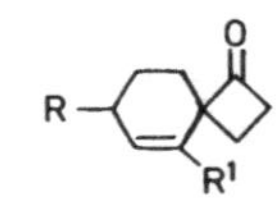	
9	R = CH_3, R^1 = H	*30*	X = PhS	D	(61)[c]	75)
10	R = H, R^1 = CH_3	*34*	X = CH_3O	D	(68)[c]	83)
11	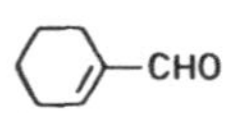	*34*	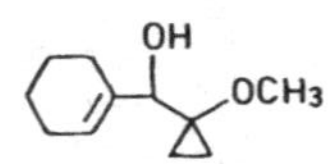	D	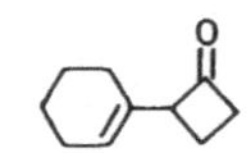 (52)[c]	80)
12	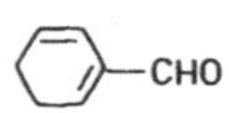	*34*	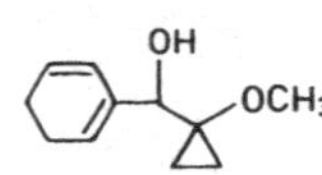	D	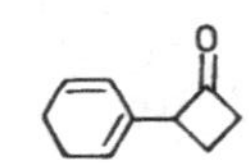 (63)[c]	83)
13	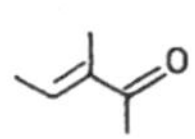	*30*	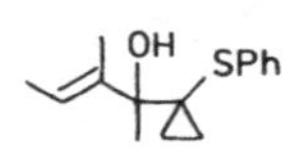 (96)	D	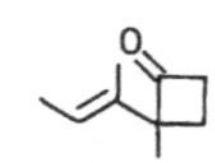 (92)	73 74)
14	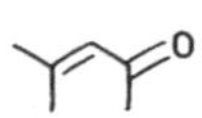	*30*	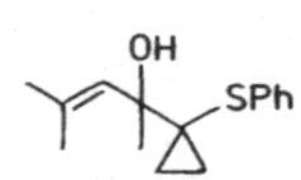 (89)	D	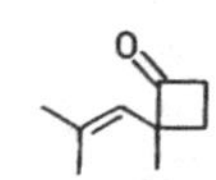 (47)	73,74,82)

Table 5. (continued)

Entry	Aldehyde or ketone	Reagent	Adduct (% Yield)	Rearrangement method	Cyclobutanone (% Yield)	Ref.
15	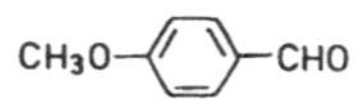	*34*	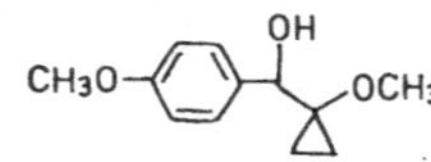	D	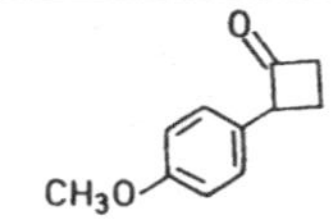 (64)[c]	83)
16	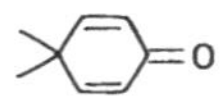	*30*	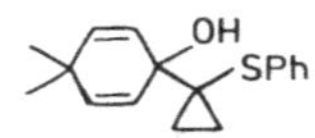 (U)[b]			82)
	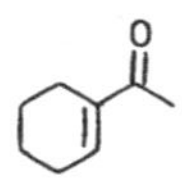		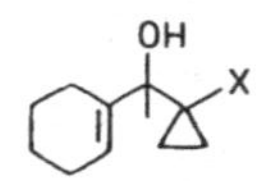 (88)		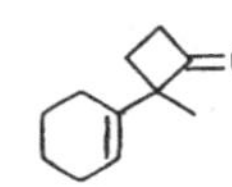	
17		*30*	X = PhS	D	(83)	73 74)
18		*34*	X = CH_3O	D	(82)	83)
19	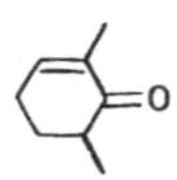	*30*	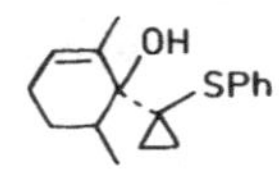 (88)	C	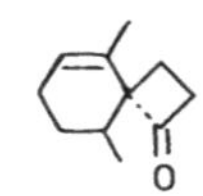 (88)	73 74)
20	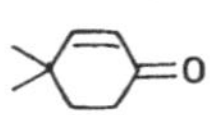	*30*	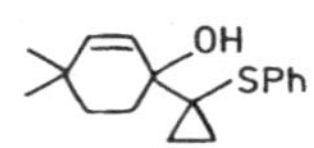 (86)			73)
21	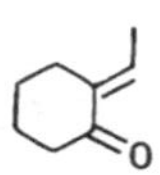	*30*	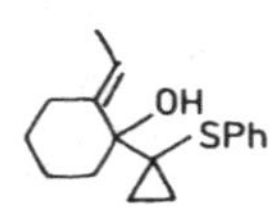 (85)	C	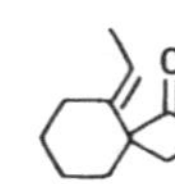 (94)	73, 74)

23	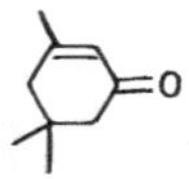	*30*	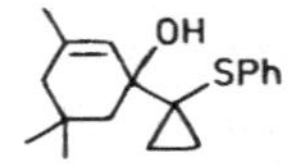			73)
24	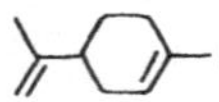	*34*	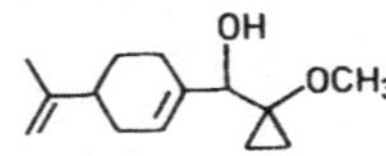		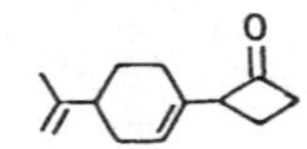 (50)^c	83)
	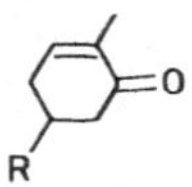		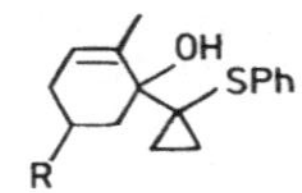			
25		*30*	R = $CH_3C(=CH_2)$— (98)			73)
26			R = iC_3H_7 (72)			
	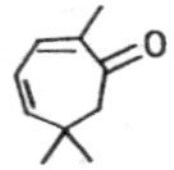		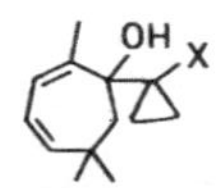		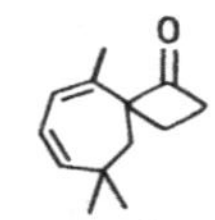	
27		*30*	X = PhS	F	(13)^c	84)
28		*34*	X = CH_3O	D	(54)	84)
29	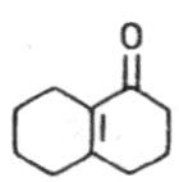	*34*	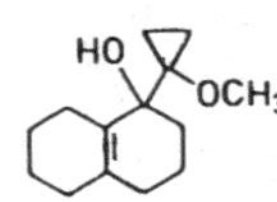	D	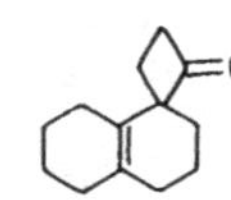 (79)^c	83)
30	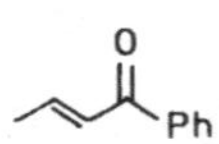	*34*	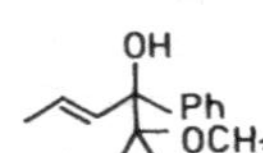	D	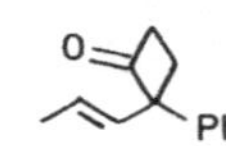 (15)^c	83)

Table 5. (continued)

Entry	Aldehyde or ketone	Reagent	Adduct (% Yield)	Rearrangement method	Cyclobutanone (% Yield)	Ref.
	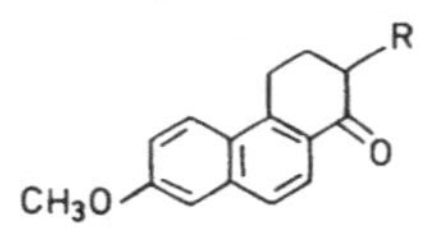		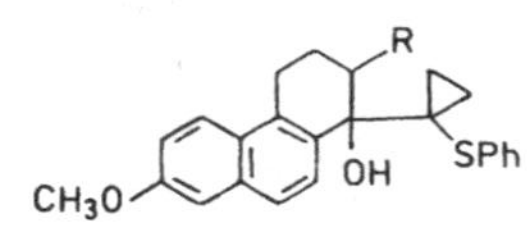			
31		*30*	R = H (87)			
32			R = CH_3 (96)			

[a] Method A: TsOH, $HgCl_2$, decalin, 70 °C; Method B: TsOH, H_2O, PhH reflux; Method C: $SnCl_4$, CH_2Cl_2 room temperature; Method D: 48% aqueous HBF_4, ether or THF; Method E: $(CH_3)_3OBF_4$, CH_2Cl_2, then aq. NaOH; Method F: HCO_2H

[b] U = undetermined yield

[c] Overall yield from starting carbonyl partner

[d] Ratio of diastereomers, only major diastereomer depicted

Cyclopropylcarbinyl-cyclobutyl ring expansions (Eq. 5) are facilitated by the presence of the heteroatom substituent in the order O > S > Se. In this case, the heteroatom stabilized cyclobutyl cation (see Eq. 39) can suffer hydrolysis to give the

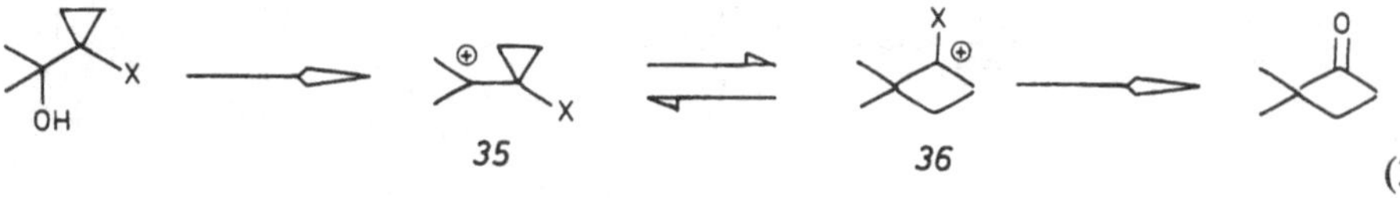

(39)

cyclobutanone. Table 5 demonstrates that solvolysis of these adducts indeed produces the desired cyclobutanones. For the sulfur analogue, the efficiency of this process is related, in part, to the relative stability of *35* and *36* (Eq. 39). As *35* becomes more stable, as in the case of aryl or vinyl substituents, the phenylthio by-products become effective scavengers of *35* thereby lowering yields. Thus, best yields with few interfering reactions using *30* occur with non-conjugated aldehydes and ketones, with the former somewhat better. Conjugated aldehydes and ketones can benefit by the presence of a thiophenol scavenger such as acrolein [78]. Use of the oxygen reagent *34* is not complicated by the phenylthio by-product of the sulfur reagent *30*; however, its lower accessibility makes it less attractive. A slight variation in which the alkoxide is derivatized in situ to make the oxygen a better leaving group such as a phosphite may also be used advantageously (Eq. 40) [74].

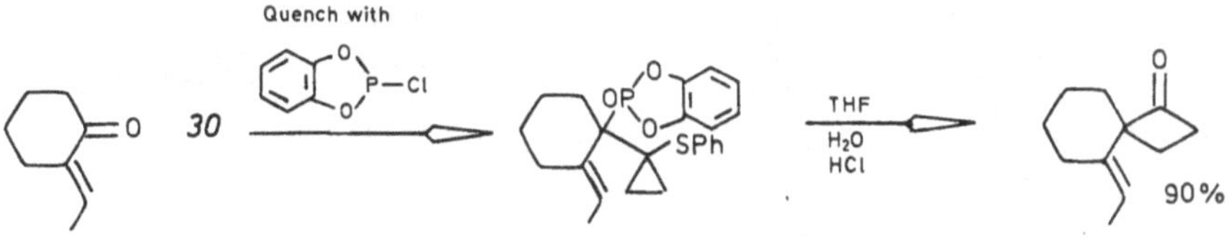

(40)

Two key aspects of this annulation differentiate it from the cyclopropylide. First, conjugated aldehydes and ketones undergo carbonyl addition and consequently cyclobutanone annulation with *30* and *34* [74] but suffer conjugate addition and spiropentane formation with the ylide (Eq. 28) [30]. Second, complementary stereochemistry normally occurs. Whereas, the ylide from *9* produces the cyclobutanone with the carbon-carbonyl bond on the more hindered face of the starting substrate; the annulation via *30* produces the opposite stereochemistry. The latter probably derives from thermodynamic considerations in which the equilibration between *35* and *36* places the heteroatom bearing cyclobutyl carbon in the sterically least crowded region. If equilibration between *35* and *36* is not rapid relative to hydrolysis of *36*, then the stereochemistry reflective of preferential stereochemistry of migration may begin to emerge. As shown in Tab. 5, entry B-28, the stereochemistry of the product is dependent on the nature of the rearrangement conditions. The efficiency of this approach is highlighted by the smooth participation of otherwise recalcitrant carbonyl partners. For example, the ketone of entry B-35 is claimed to fail to react with olefination reagents [79].

Rearrangement under anhydrous conditions serves as an entry to cyclobutenes. The two most commonly employed reagents are TsOH (Eq. 41) and the Burgess reagent [$CH_3O_2CNSO_2N(C_2H_5)_3$] (Eq. 42), both in refluxing benzene [74]. Hydrolysis of these enolthioethers converts these reactions into another method for cyclobutanone formation.

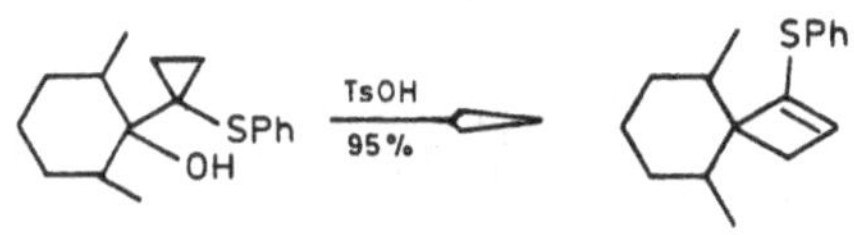

(41)

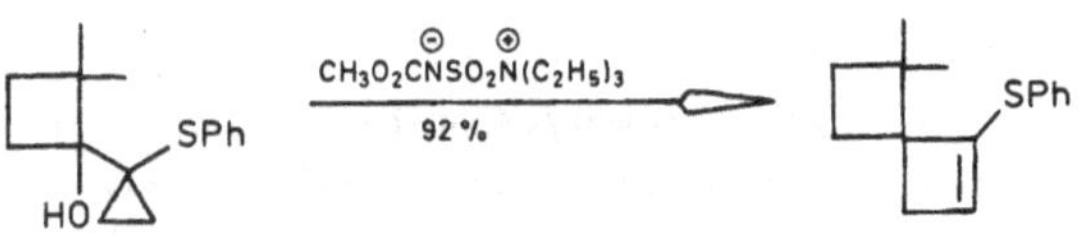

(42)

Dehydration without rearrangement occurs regioselectively with thionyl chloride in pyridine (Eq. 43). The selenium analogues have shown similar behavior (Eq. 44) [87]

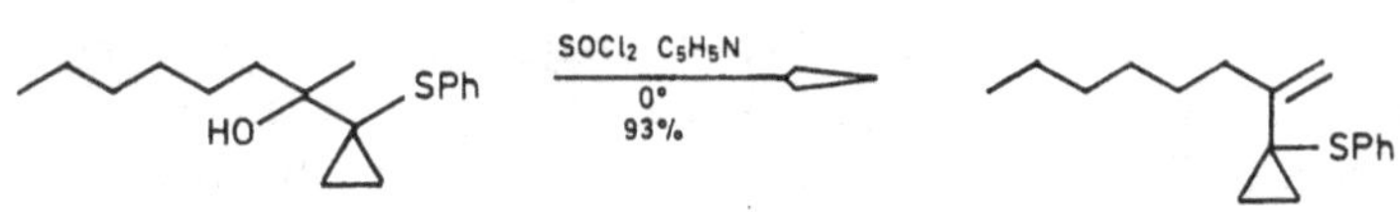

(43)

although higher temperatures appear to be required. The propensity for β-

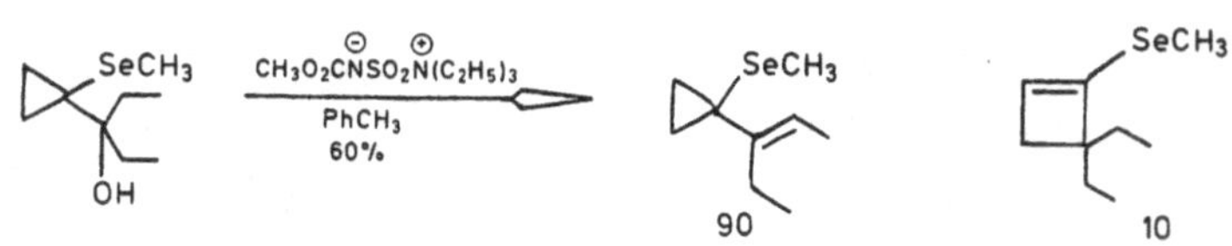

(44)

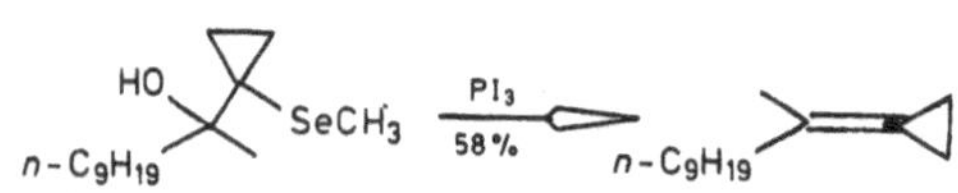

(45)

hydroxyselenides to undergo vicinal elimination to olefins also converts such adducts to alkylidenecyclopropanes (Eq. 45) [77].

7 Electrophilic Cyclopropyl Heteroatom Substituted Conjunctive Reagents

Inverting the electronic sense of the small ring conjunctive reagent expands the range of substrates that would be suitable reaction partners. The aldehydes *37a* [88] and *38* [89] are available by the formylation of the corresponding lithium salt with DMF.

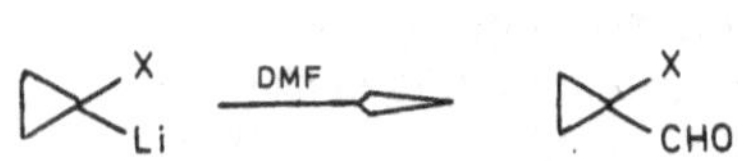

37 X = SAr *a)* Ar = Ph
b) Ar = 2,6-$(CH_3O)_2C_6H_3$
38 *a)* X = $SeCH_3$ *b)* X = SePh
39 *a)* X = OTHP *b)* X = OSi

(46)

Alternatively, *37a* and *37b* derive from alkylation of the arylthioacetonitrile followed by reduction of the nitrile directly to the carboxaldehyde (*37a* in 71%[90]; *37b* in

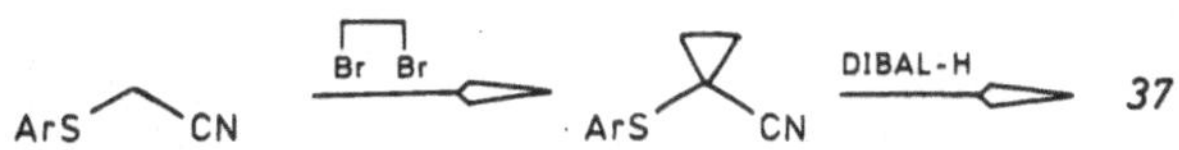

(46a)

47% overall[88], Eq. 46). The bifunctional aldehyde *42*, which exists as its lactol *43*, arises by the direct formylation method since the lithium salt *41* is nicely available by direct metallation of the cyclopropane *40* (Eq. 47)[92]. Acylation of 1-lithiumcyclopropylphenyl sulfide *30* with methyl benzoate to give *44* has also been recorded (Eq. 48)[92].

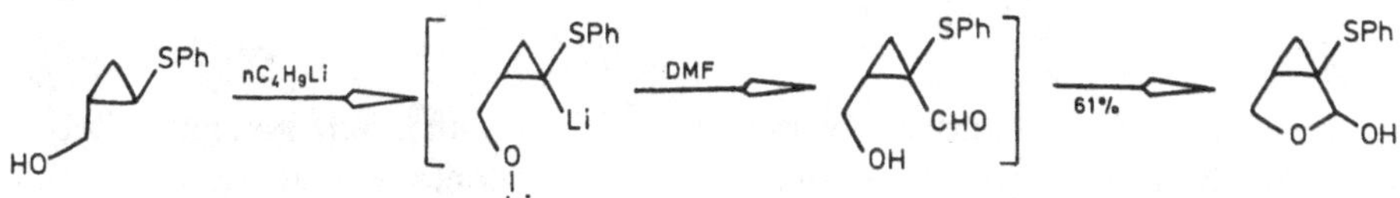

40 41 42 43

(47)

(48)

The oxygen derivative *39a* has its origins in the acyloin reduction product of diethyl succinate *45*. A benzylic acid type of rearrangement of in situ formed

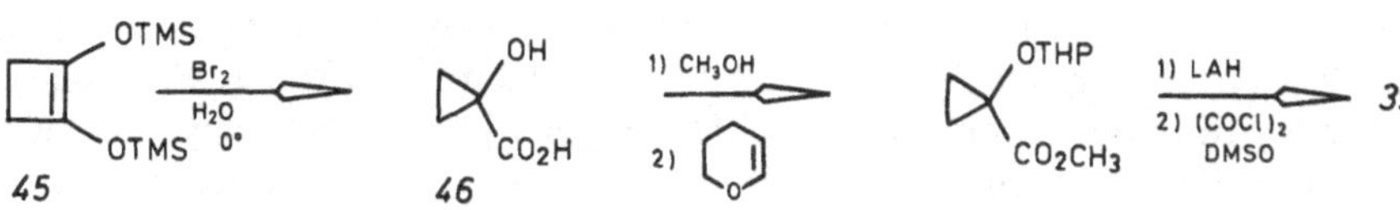

(49)

cyclobutane-1,2-dione creates the requisite cyclopropyl derivative *46*. Standard procedures complete the sequence as summarized in Eq. 49[93].

These electrophilic conjunctive reagents require donor reactants. The cyclopropylcarbinols as precursors to cyclobutanones arise by simple addition of organometallics. For example, the cyclobutanone *47* derives by addition of vinyllithium to *44* followed by rearrangement with aqueous fluoroboric acid[92]. In some cases, this route to cyclopropylcarbinols is preferred. Addition of *41* to aldehydes or ketones

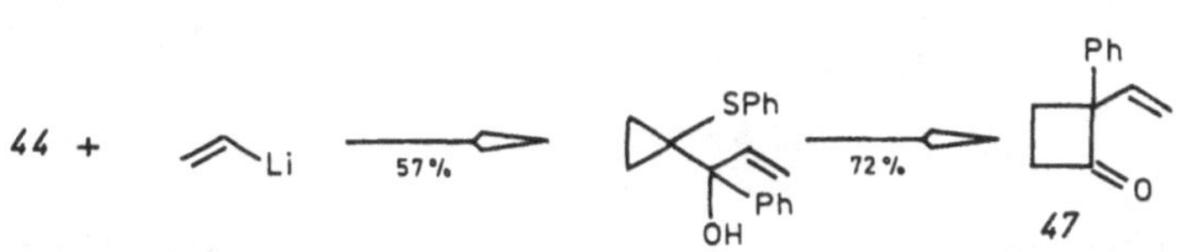

(50)

causes substantial enolization, presumably a reflection of the presence of the internal alkoxide base[94]. On the other hand, addition of Grignard conjunctive reagents to

43 followed by chemoselective acylation of the primary alcohol with pivaloyl chloride (Eq. 51) gives good yields of the desired cyclopropylcarbinols 48 [95].

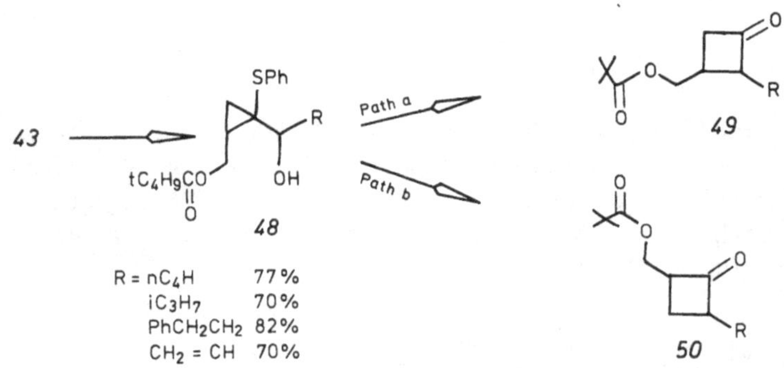

(51)

Rearrangement of these adducts proceeds smoothest with triethyloxonium fluoroborate. While initial expectations that the cyclobutanones should posses the 2.3-disubstitution pattern as in *49* by migration of the more substituted cyclopropyl carbon (i.e. bond "a" as in Path a), the exclusive product arises from migration of the less substituted carbon (bond "b" as in Path b). This unusual regioselectivity may result from the fact that an acyloxymethyl group is inductively electron withdrawing — an effect which would destabilize a transition state developing positive charge on the neighboring carbon as in Path "a". The utility of such remote inductive effects for control of selectivity would appear to be greater than is generally assumed and provides another dimension for selectivity in the design of new conjunctive reagents.

Enolates also serve as suitable reaction partners in a directed aldol condensation (Eq. 52) [88] Dehydration of the aldol *51* to give enone ($POCl_3$, HMPA then C_5H_5N) followed by reduction [$LiAlH(nC_4H_9)(iC_4H_9)_2$] produces a vinylogue of a cyclopropylcarbinol *53*. Aqueous fluoroboric acid smoothly rearranges *53* to the

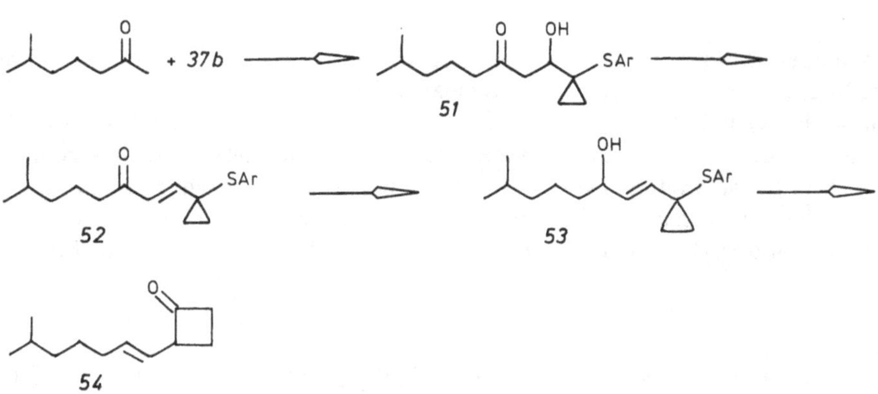

(52)

vinylcyclobutanone *54*. The use of the 2,6-dimethoxyphenylthio group is important in improving the efficiency of this sequence by

1) enhancing the rate of rearrangement due to the relief of steric congestion and the electron releasing effects of the oxygen substituents and

2) decreasing the trapping of carbonium ion intermediates by the thiol by-product by sterically shielding the sulfur.
This four step operation to convert ketones to vinylcyclobutanones (Eq. 53)

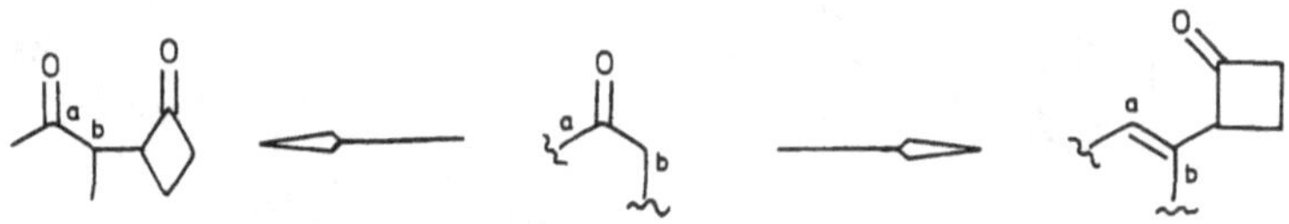

(53)

proceeds so well that no purification is necessary at any step except at the end. For the example of Eq. 52, the overall yield from starting ketone and the electrophilic cyclopropyl conjunctive reagent is 44%.

In the oxygen analogue, the enone itself is sufficiently electrophilic that attempted hydrolysis of the THP ether in enone *55* only produces the ring expanded cyclobutanone (eq. 54) [96]. The overall result of this sequence for electrophilic partners constitutes a 2-oxocyclobutylation alpha to a carbonyl group (see Eq. 53).

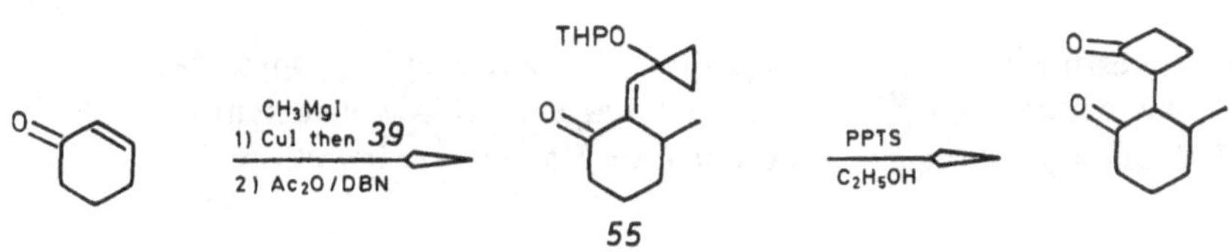

(54)

Olefination reactions converts these electrophilic conjunctive reagents into vinylcyclopropanes (Eq. *55–58*) [89,91,93]. While in the simple aldehydes, the normal

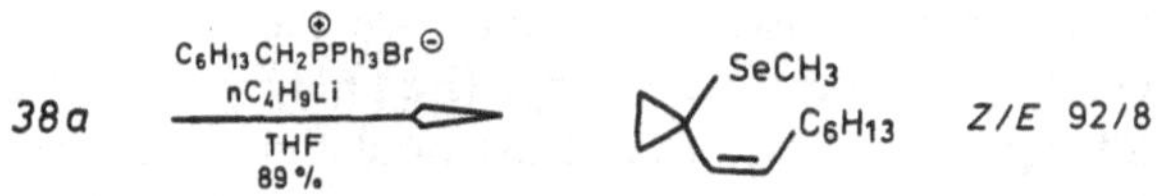

(55)

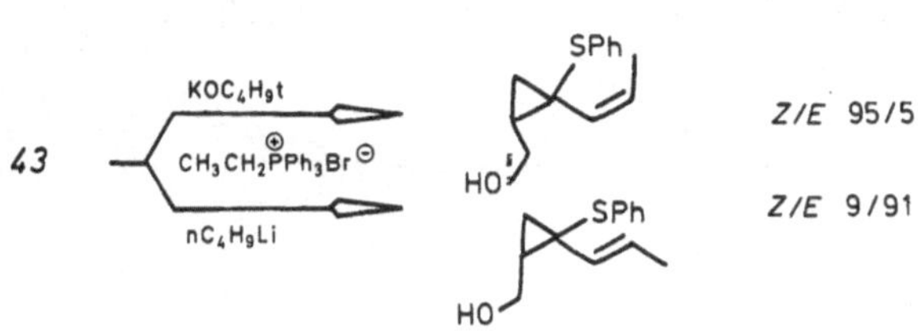

(56)

(57)

preference for the Z geometry is observed, olefin geometry in the latter case depends upon the chosen base [91]. Peterson olefination (Eq. 57) gives a stereoisomeric mixture of olefins [93]. Metalated phosphine oxide serves as an olefination conjunctive reagent (Eq. 58) in which the *E* olefin greatly dominated (>97%) (see Eq. 58) [97].

43 +

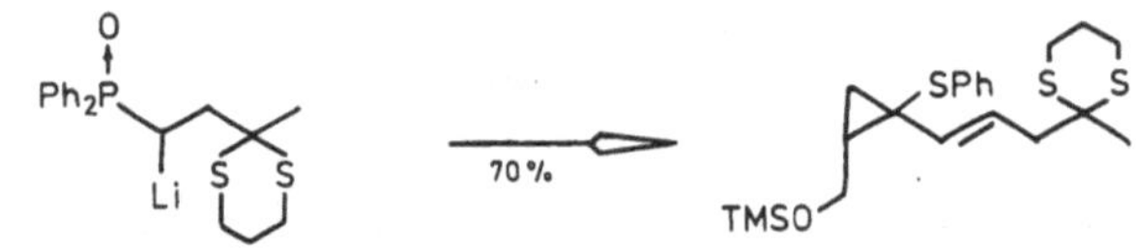

(58)

A bis-vinylogue of a cyclopropylcarbinol *57* arises by standard transformations (Eq. 59) from the Peterson olefination product *56*. Here too, acid induced rearrangement proceeds exclusively to the cyclobutanone and not to larger ring products [93].

56

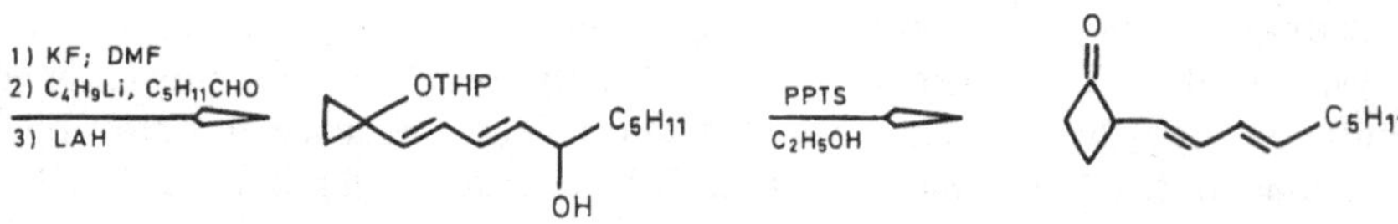

(59)

Another interesting class of electrophilic conjunctive cyclopropyl reagents are cyclopropanone hemiacetals [98]. The parent reagent *58* is conveniently available from ethyl β-chloropropionate as outlined in Eq. *60* [99]. Its utility stems from

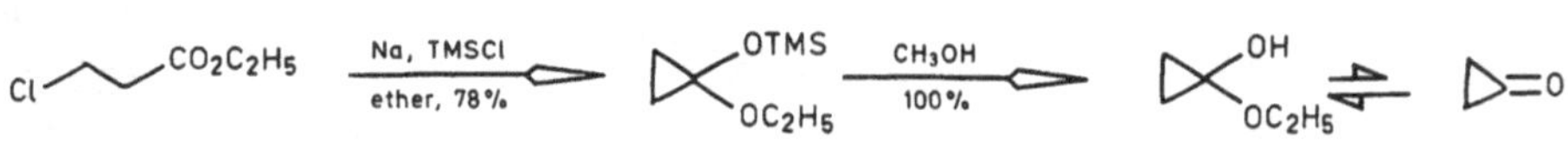

58

(60)

the high reactivity of cyclopropanone which is in dynamic equilibrium with the hemiacetal in solution. The most useful reactions are the addition of vinyl and acetylenic organometallics to generate the vinyl (Eq. 61) [99b, 100, 101] or propargyl (Eq. 62) [98a, 102] cyclopropanols. Acid induces cyclobutanone formation (Eq. 61a) [101b]. The propargyl cyclopropanols permit access to either the *E* (Eq. 61a) [103] or *Z* (Eq. 61c) [98a] vinyl cyclopropanols. A related approach to vinylcyclopropanols

58

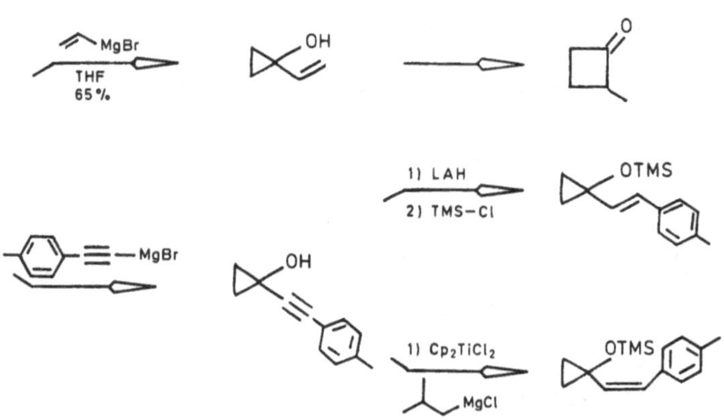

(61a)

(61b)

(61c)

utilizes a 1,3-reductive elimination of the carbonyl adducts of 1,3-dichloroacetone as shown in Eq. 62 [104].

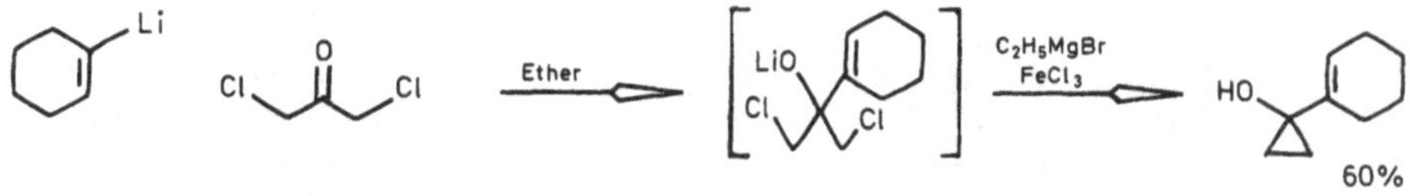

(62)

8 Substitutive Spironannulation

The juxtaposition of functionality in a vinylcyclopropanol creates a possibility for unusual olefin reactivity. A relationship of *60* to an enol *59* is evident in that a cyclopropyl ring is inserted between the two functionalities of the enol. The reactivity of an enol stems from delocalization of the electron lone pairs on oxygen into the olefin to increase olefinic nucleophilicity. Since the bonding electrons of the C—C cyclopropyl bonds are polarizable, enhanced olefin nucleophilicity may still emerge from *60* where the lone pair delocalization from oxygen to the olefin is mediated by the cyclopropyl ring. The resulting composite functional group would react with electrophiles with participation of the ring as illustrated in Eq. 63. A corollary for such a bonding would be enhanced olefin reactivity.

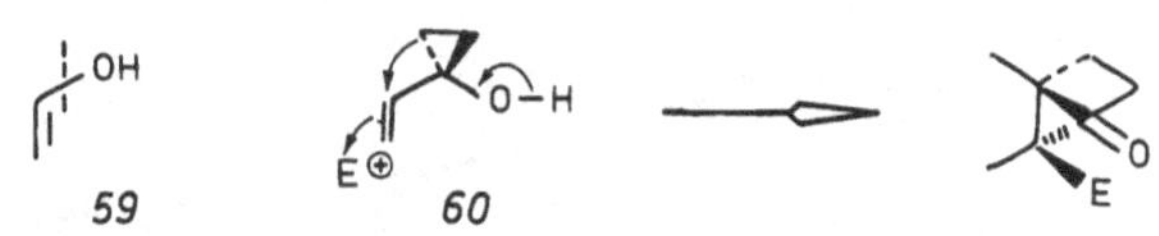

(63)

Typical electrophiles that attack olefins like Br^+ (Eq. 64) [105] and $(OH)^+$ (Eq. 65) [105] do lead to electrophilic addition with ring enlargement [101 a]. Most interesting is the ability for electrophiles that normally do not attack olefins to also react. Thus, an oxycarbonium ion generated from an acetal smoothly alkylates the double bond of this composite functional group in an overall highly stereocontrolled 1,1,2-trialkylation of a simple ketone as illustrated in Eq. 66 [106]. The chemo- and

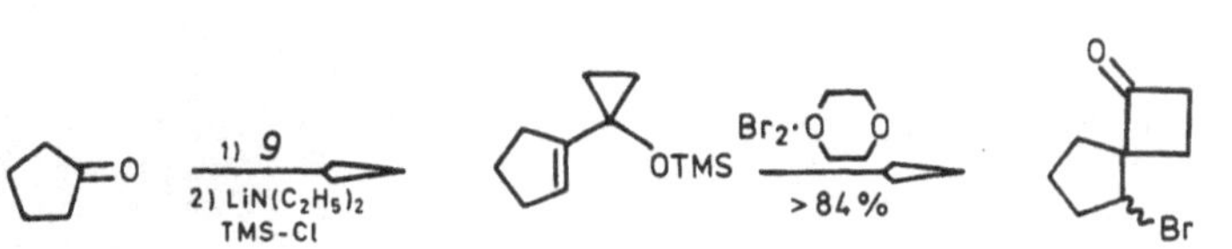

(64)

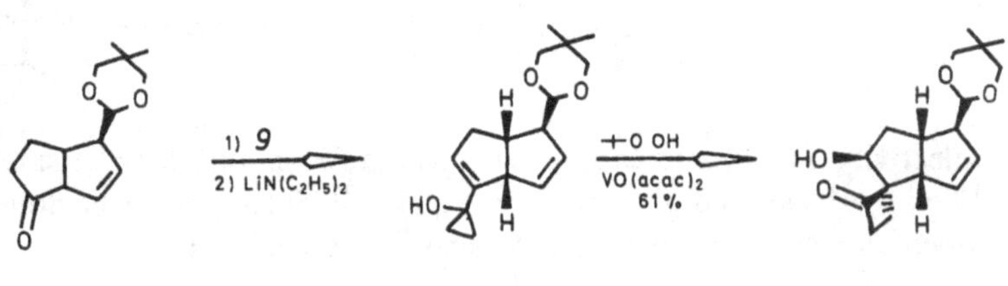

(65)

regioselectivity inherent in this scheme permits a cyclization as illustrated in Eq. 67.

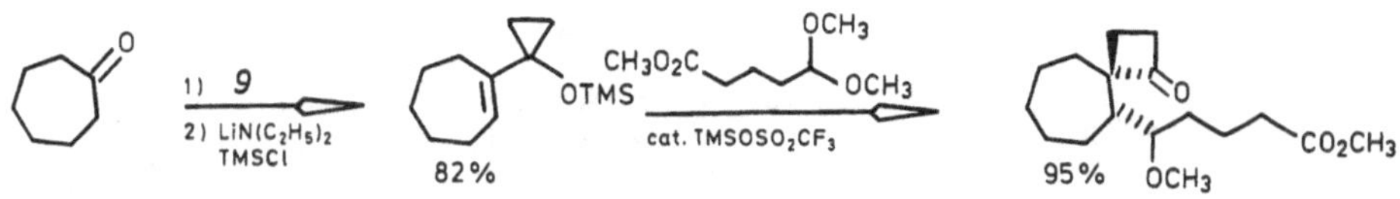

(66)

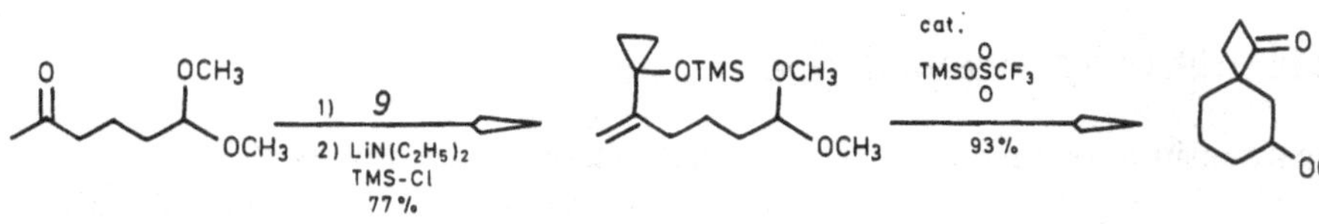

(67)

8.1 γ-Butyrolactone Annulation

Cyclobutanone annulation onto a carbonyl group translates into γ-butyrolactone annulation because of the facility of the Baeyer-Villiger reaction (Eq. 68a) [8]. Indeed, the reaction proceeds sufficiently rapidly that even basic hydrogen peroxide effects the oxidation; whereas, with less reactive carbonyl partners, peracids must be used.

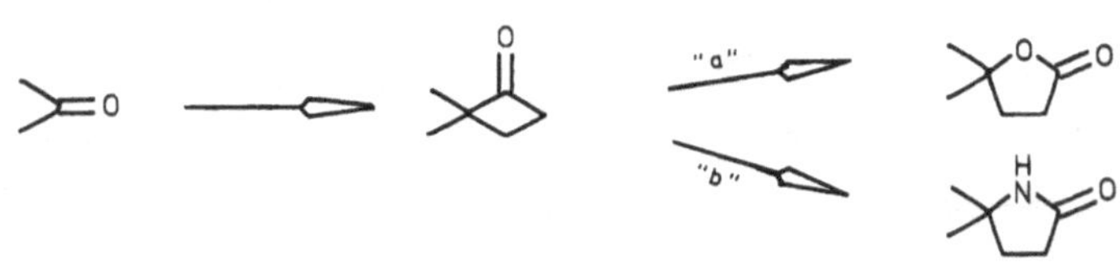

(68)

Perbenzimidic acid also effects rearrangement [35]. Use of basic hydrogen peroxide permits the substrates to contain olefins with immunity as in Eq. 69. The high

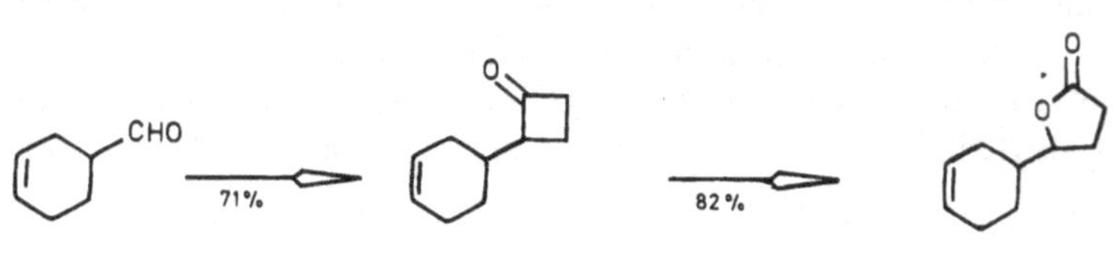

(69)

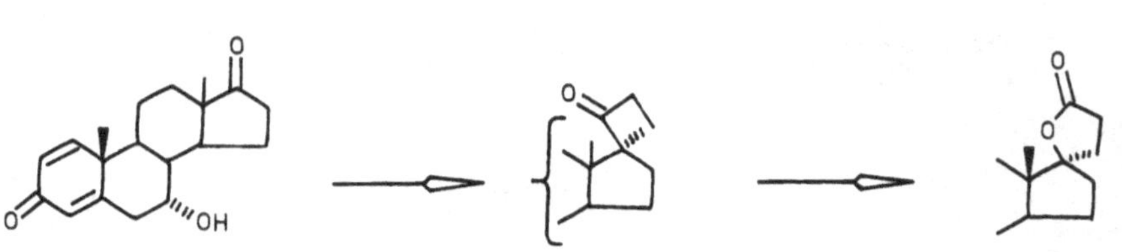

(70)

chemoselectivity and diastereoselectivity available in this lactone annulation are examplified by the steroid dienone as in Eq. 70 [53]. Since γ-butyrolactones rearrange to cyclopentenones in the presence of acid [107], the overall process as outlined in

Eq. 71 becomes an effective cyclopentenone fusion. Such a sequence has proven to be the preferred preparation of polyquinanes exemplified in Eq. 72 [52].

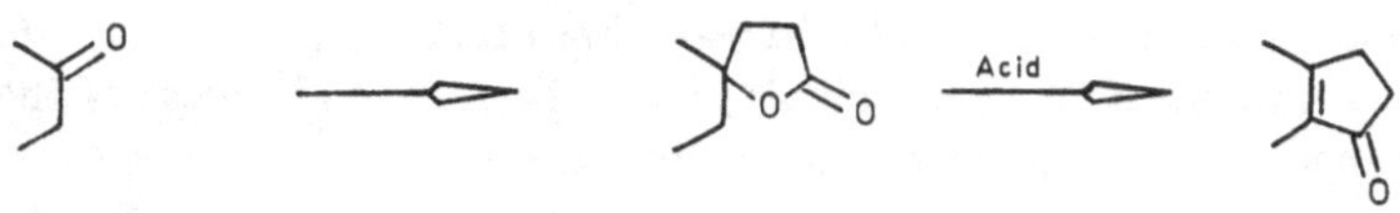

(71)

Beckmann like rearrangements have not been extensively explored but should translate this spiroannulation into a γ-butyrolactam synthesis [108].

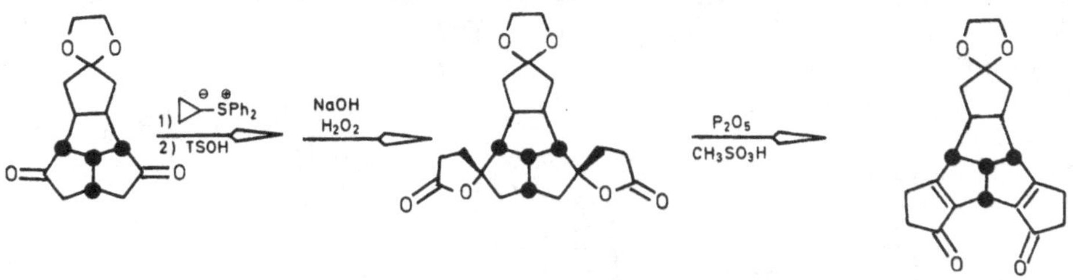

(72)

8.2 1,1-Cyclopentanone Annulation

Whereas lactone annulation invokes a relief of strain of the four membered ring by migration of the ring bond to an electron deficient oxygen, a similar migration to an election deficient carbon creates a cyclopentanone synthesis (Eq. 73). The release of approximately 84 kJ/mole (20 kcal/mole) provides a strong driving force. Thus, the 1,1-cyclobutanone annulation of ketones translates into a 1,1-cyclopentanone annulation.

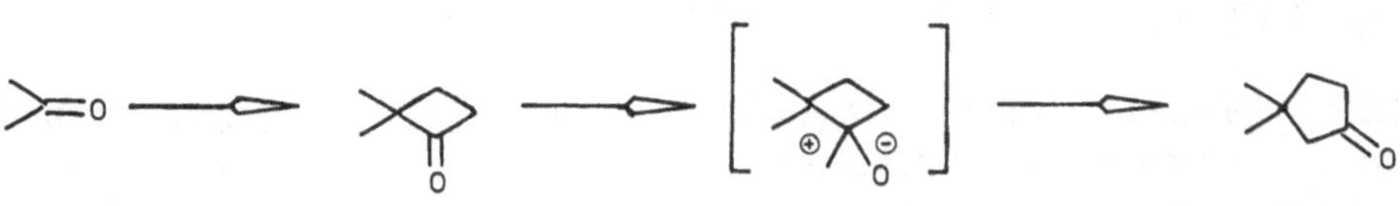

(73)

For example, an oxaspirohexane *62*, readily available by condensing cyclobutanone *61* with dimethylsulfonium methylide, rapidly rearranges (isomerizes) to the cyclopentanone *63* upon exposure to a catalytic amount of lithium bromide [55]. The high diastereoselectivity of the initial cyclobutanone formation translates into a high diastereoselectivity for cyclopentanone annulation as this example of Eq. 74 demonstrates.

Any method of ring expansion such as the use of diazoalkanes [7,109] or sulfur stabilized carbanions as exemplified in Eq. 75 [6b] makes this approach very flexible with respect to the control of substitution on the ring. The regioselectivity appears determined by migration of the carbon that best bears a positive charge which is the more substituted carbon. As noted in Eq. 32, the selenoxide becomes a good leaving group when adjacent to a neighboring group that stabilizes a carbonium ion — in this case a cyclopropanol whose effectiveness stems, in part, from the release of strain [37]. The high strain associated with a cyclobutyl ring also makes a corresponding cyclobutanol ring an effective neighboring group [110]. An attractive approach to 1,1-cyclopentanone annulation utilizes selenoxide stabilized carbanions as in Eq. 76. Similar to Eq. 32, the intermediate selenoxide *64* cannot be detected because of the facility of the rearrangement due to the release of ring strain.

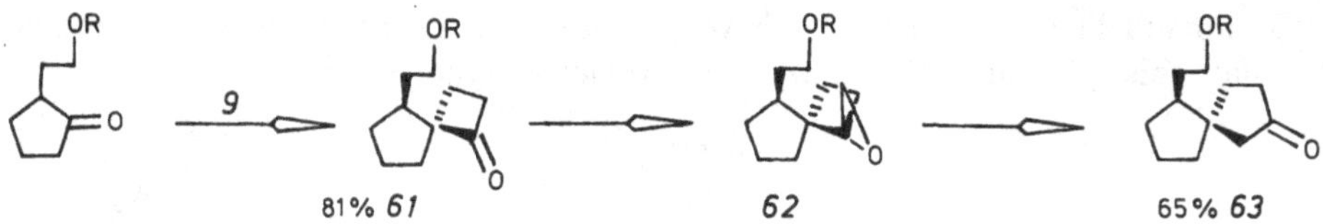

(74)

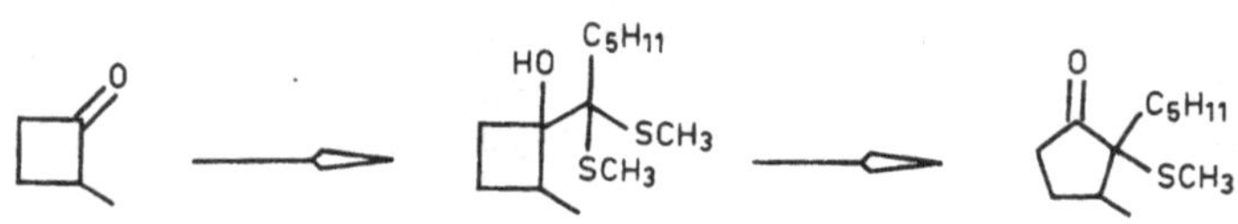

(75)

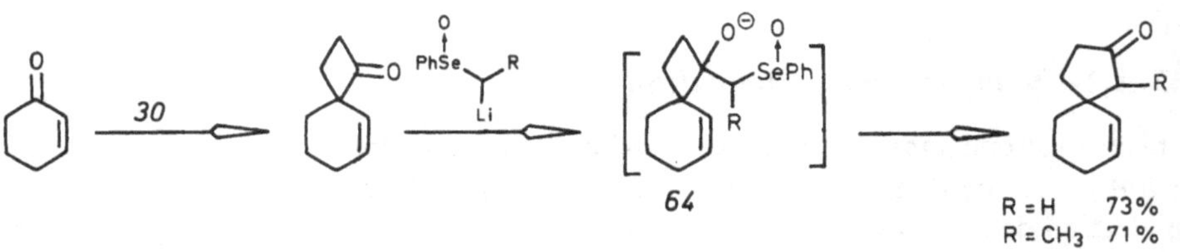

(76)

8.3 1,2-Cyclopentenone Annulation via Cyclobutanones

Placing the electron deficient migration terminus within the original carbonyl partner converts the 1,1-cyclobutanone annulation into a 1,2- or lateral cyclopentenone annulation as summarized in Eq. 77. In the ring enlargement of *65* to *66*,

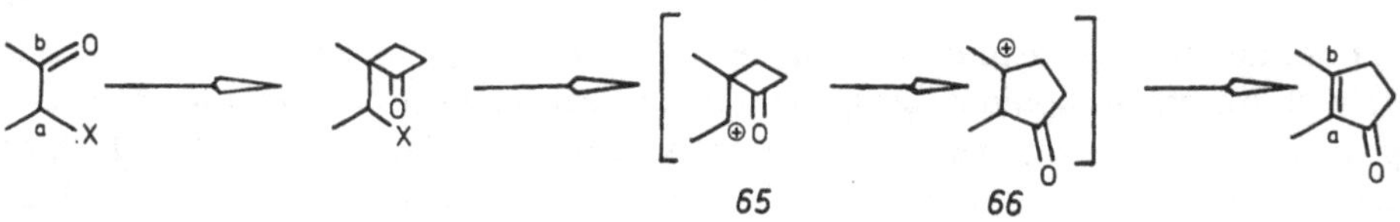

(77)

the group that best stabilizes a positive charge which should preferentially migrate is the carbonyl group since it is the equivalent of an acylium ion. The overall annulation attaches the carbonyl carbon of the newly annulated ring onto the alpha carbon [C(a)] of the starting ketone.

Not too many examples of this process exist. Protonation of the vinyl cyclobutanones derived from α,β-unsaturated ketones creates just such a reactive intermediate.

Indeed, exposure of the vinylcyclobutanone *67* to acid produces the perhydroazulene skeleton *68* as summarized in Eq. 78 [112]. With more ready access to the types of substituted cyclobutanones required for this lateral cyclopentannulation (see substitutive spiroannulation), milder conditions and thus better processes will probably ultimately be developed.

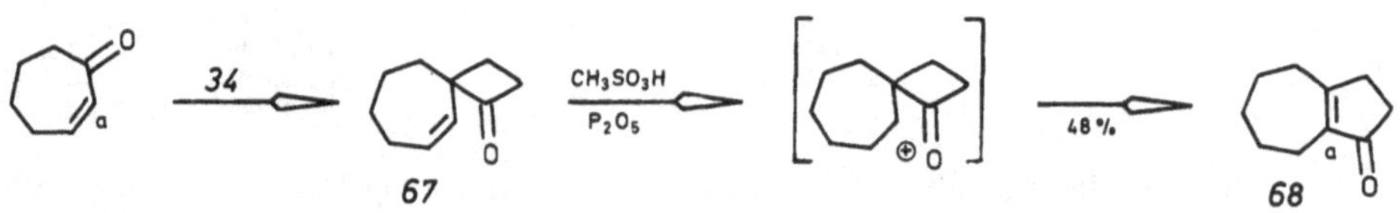

(78)

8.4 1,2-Cyclopentyl Annulations via Vinylcyclopropanes

The ready introduction of vinylcyclopropanes into organic molecules utilizing these small ring conjunctive reagents awakens interest into the previously little used vinylcyclopropane rearrangement to cyclopentenes. The facility of this reaction shows a dramatic dependence on the nature of the substitution. In particular, a vinylcyclopropanol shows an acceleration compared to the parent ($\Delta\Delta G^{\ddagger}_{350\,°C}$ 21 kJ/mole (5 kcal/mole)) [12]. The first applications take advantage of the base catalyzed opening of oxaspiropentanes [62]. As noted in Eq. 34, either regioisomeric vinylcyclopropane *69* and *70* is available by appropriate choice of base [49]. Thermolysis at 330 °C smoothly rearranges to the enol silyl ethers *71* and *72* respectively. Hydrolysis completes the 1,2-cyclopentanone annulation in 58–66 % overall yields from the starting ketone (Eq. 79). The regioselectivity of the base opening translates into the

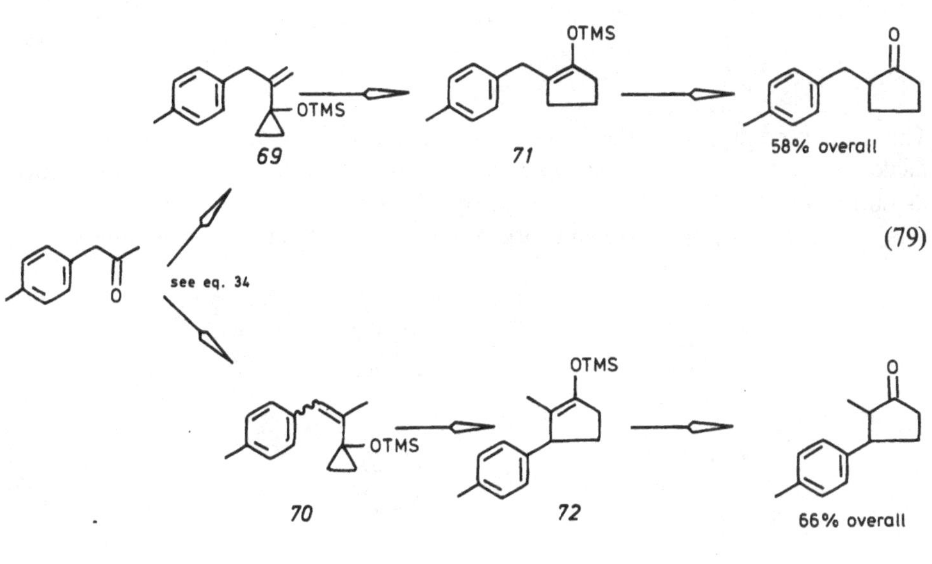

(79)

substitution pattern of the final cyclopentanone. The diastereoselectivity with 2-methylcycloalkanones is low (Eq. 80) [63].

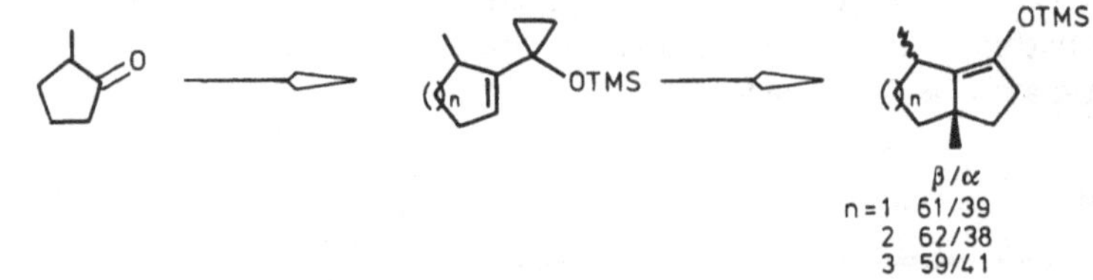

(80)

The intermediate enol silyl ether permits further regioselective substitutions such as bromination followed by dehydrobromination (Eq. 81) [49] and alkylation (Eqs. 82 [93] and 83 [103]). Thus, in addition to activating the rearrangement, the oxygen substituent regioselectivity creates an enol silyl ether, a powerful enolate synthon.

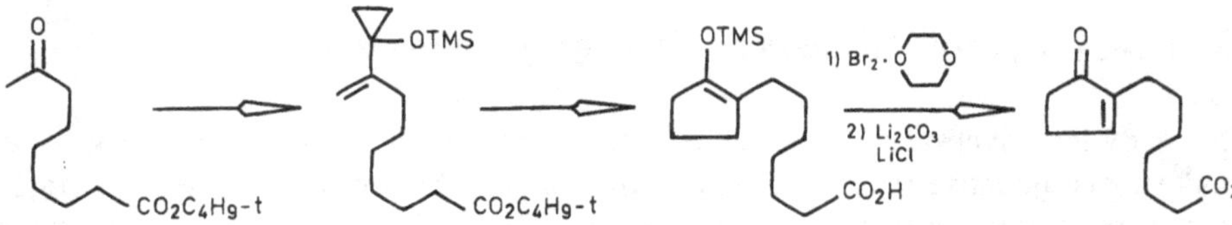

(81)

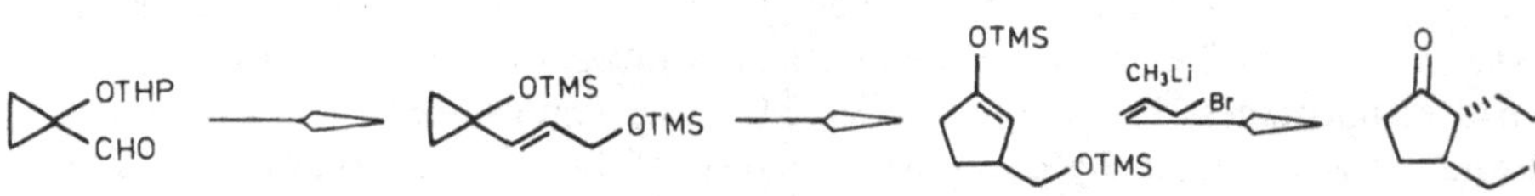

(82)

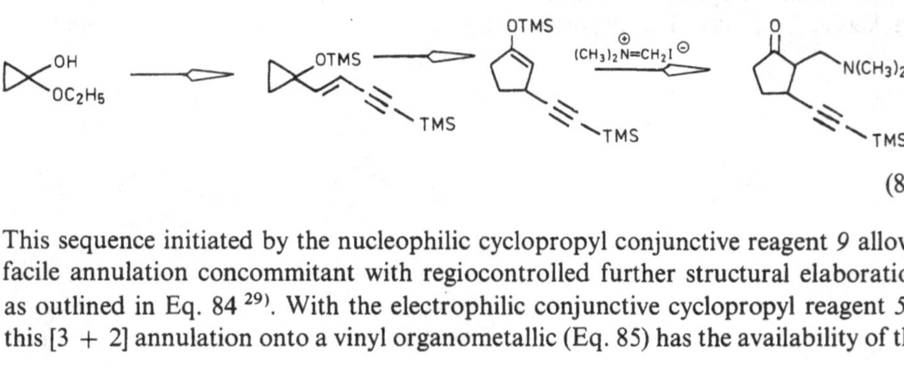

(83)

This sequence initiated by the nucleophilic cyclopropyl conjunctive reagent *9* allows facile annulation concommitant with regiocontrolled further structural elaboration as outlined in Eq. 84 [29]. With the electrophilic conjunctive cyclopropyl reagent *58*, this [3 + 2] annulation onto a vinyl organometallic (Eq. 85) has the availability of the

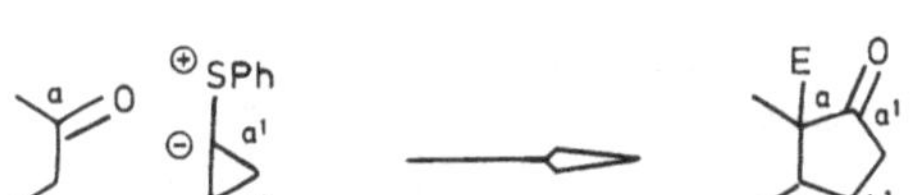

(84)

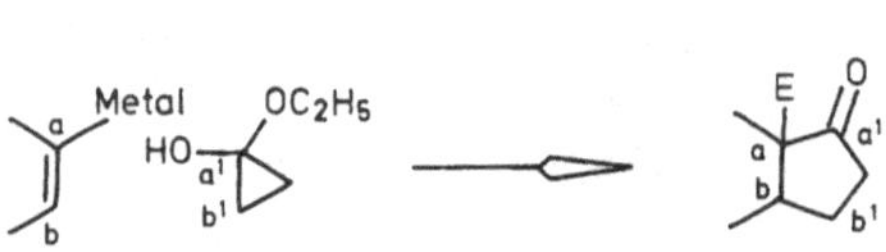

(85)

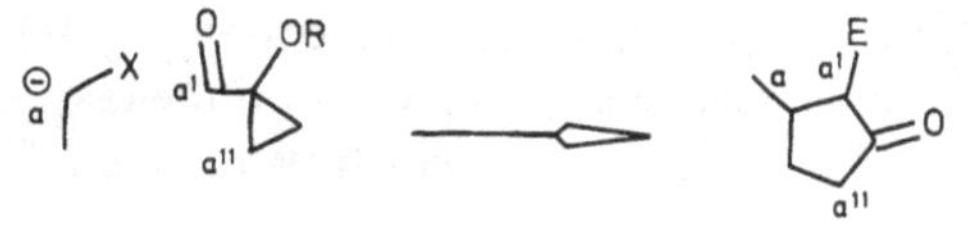

(86)

organometallic as its main limitation. The use of cyclopropylcarboxaldehyde *39* becomes a [4 + 1] annulation onto an olefination agent (Eq. 86).

Other cyclopropyl conjunctive reagents are also employed in such annulations including the parent (see Eq. 18). The closest in producing a similar structural change is the thio analogue. Some rate enhancement derives from the sulfur activation. As an annulation onto ketones, a sequence employing *37a* accomplishes the same structural change (Eq. 87a) when the annulated product *73* is hydrolyzed ($TiCl_4$,

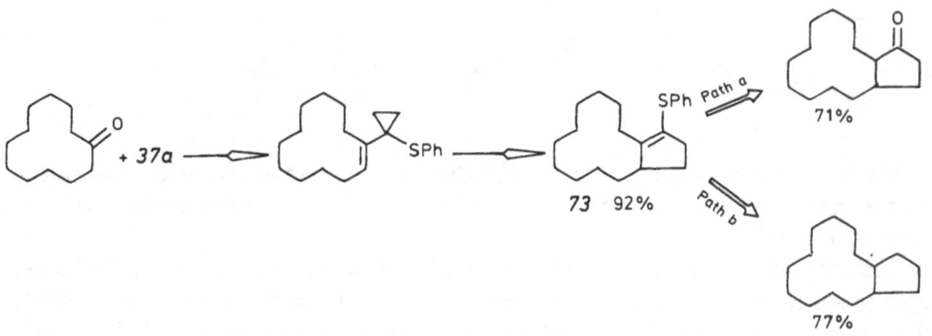

(87)

HOAc, H_2O, room temperature)[113]. Alternatively, a simple cyclopentannulation results upon reduction of *73* (Eq. 87b, W-2 Raney-Ni, C_2H_5OH, reflux)[114].

Whereas, the sulfur ylide *9* undergoes conjugate addition to enones thereby precluding 1,2-cyclopentyl annulation, the organolithium *37a* undergoes clean carbonyl addition. Cyclopentenone participates smoothly in this sequence (Eq. 88)[81].

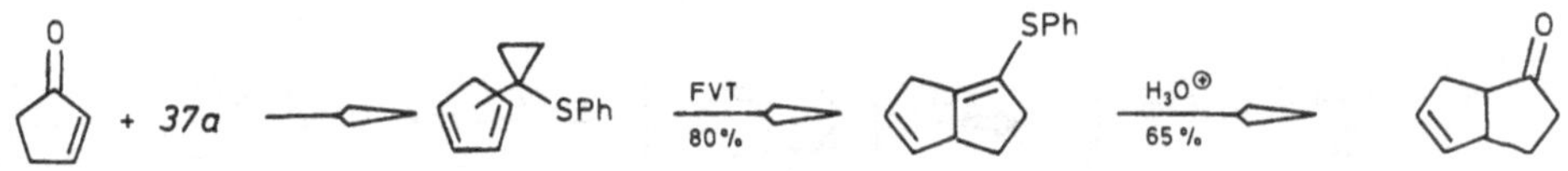

(88)

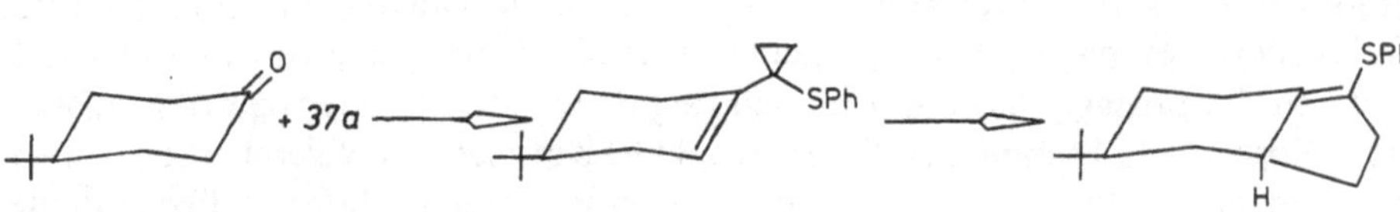

(89)

Good diastereoselectivity is observed with respect to the migration terminus (Eq. 89)[113].

The silicon analogue *74* [115] also appears to be a potentially useful conjunctive reagent in this sequence even though silicon appears to retard the rearrangement [71]. It is generated by silylation of *37a* followed by reductive lithiation (Eq. 90). The

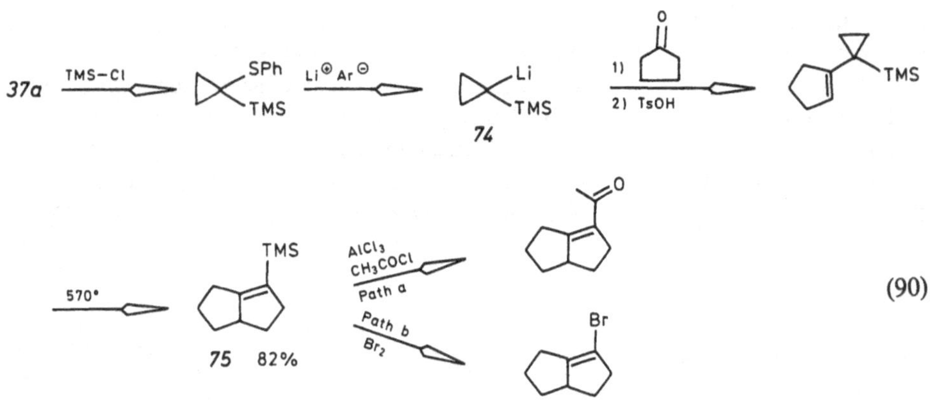

(90)

utility of the silicon analogue stems from the regioselective electrophilic substitution of the resultant vinylsilane by either carbon (Eq. 90a) or non-carbon (Eq. 90b) electrophiles [115, 116].

Heterocyclic syntheses take advantage of a heteroatom version of the vinylcyclopropane rearrangement — an iminocyclopropane rearrangement. For example, the aminocyclopropanol *76*, available analogously to the hemiketal *58*, readily

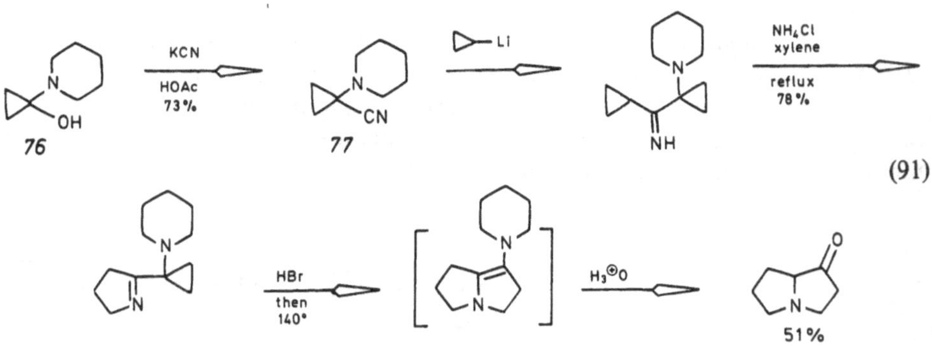

(91)

substitutes the hydroxyl group for cyanide [117]. Addition of an organolithium produces the iminocyclopropane. In this particular example, a bis(cyclopropyl) derivative is produced. Surprisingly, the unsubstituted ring preferentially underwent ring enlargement. Normally, a 1-amino substituent would be expected to accelerate the reaction. The mechanism of this heteroatom version may indeed be quite different. For example, thermolysis of the hydrobromide or hydrochloride salt is generally preferred. Thus, the reaction may be a nucleophilically triggered sequence. A second iminocyclopropane rearrangement completes the ring system of the pyrrolizidine alkaloids as summarized in Eq. 91.

9 Vinylcyclobutane Rearrangements

The comparable strain between a cyclobutyl and cyclopropyl ring suggests the feasibility of a vinylcyclobutane to cyclohexane rearrangement. Nevertheless, such a process appears to proceed much less facilely and is complicated by [2 + 2] cycloreversion. Thermal rearrangements of vinylcyclobutanones have been reported (Eq. 92) [118]. However, the synthetically most useful reactions appear to be the base

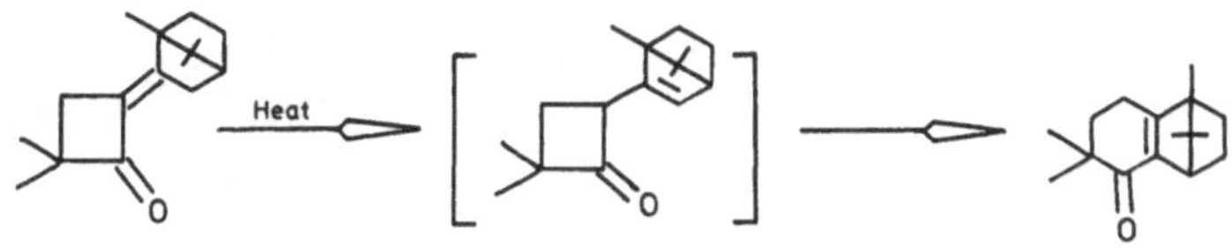

(92)

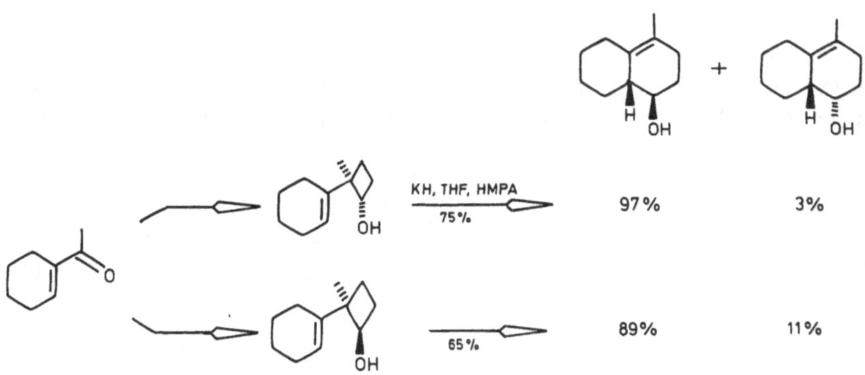

(93)

accelerated vinylcyclobutanol rearrangements (Eq. 93) [92]. Since the stereochemistry of the starting material is lost, a fragmentation readdition rationalizes this rearrangement [92, 119]. Sufficient driving force for this rearrangement exists that an aromatic ring can serve as the vinyl component (Eq. 94) [80].

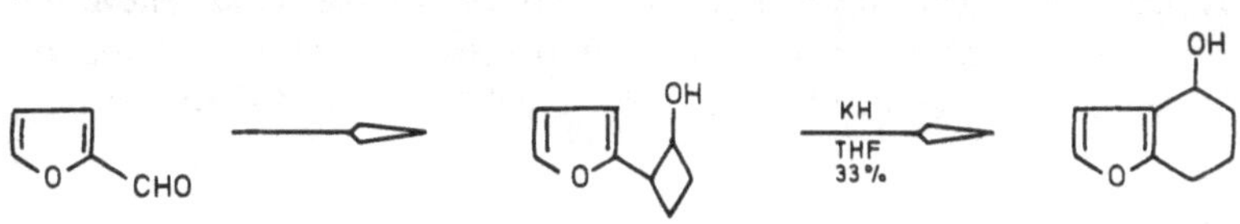

(94)

An alternative mode to release the strain of the four membered ring of vinylcyclobutanones arises from the addition of vinylorganometallics to the carbonyl group whereby a divinylcyclobutane is created [78, 120, 121]. Since equilibration of the stereoisomers *78* and *79* is fast relative to rearrangement [78], both stereoisomers nicely rearrange even though a cis divinyl orientation is required (Eq. 95).

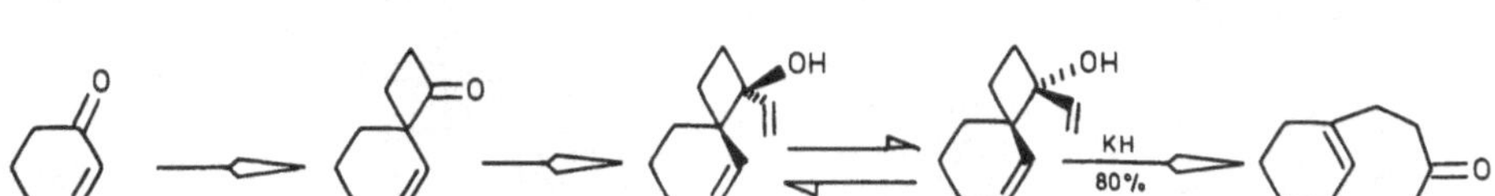

78 *79*

(95)

The vinylcyclobutane rearrangement serves as a novel [3 + 3] or [4 + 2] strategy for six membered ring formation depending upon whether a nucleophilic (Eq. 96) or electrophilic (Eq. 97) conjunctive cyclopropyl reagent is employed. The use of the

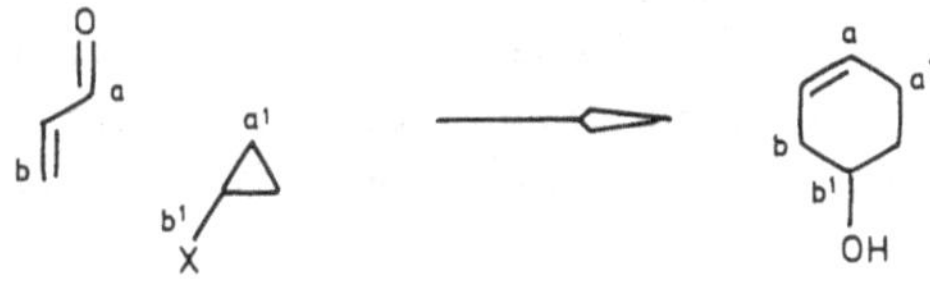
(96)

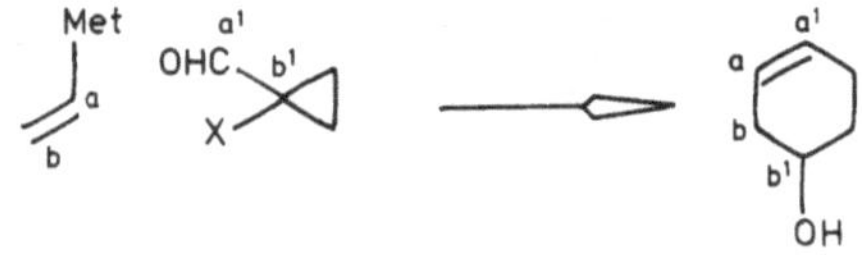
(97)

divinylcyclobutane version of the Cope rearrangement constitutes one of the best approaches to eight membered rings and formally derives from a [3 + 3 + 2] strategy (Eq. 98).

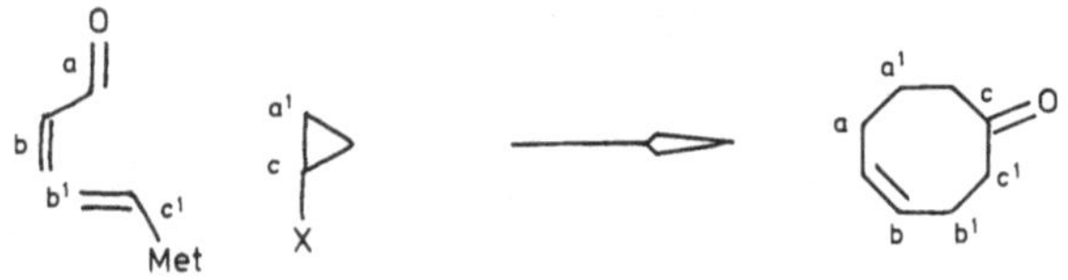
(98)

10 Formation of Acyclic Fragments by Homodiene Prototropic Shift

The vinylcyclopropenes that bear an alkyl group cis to the double bond relieve the strain of the three membered via prototropic shift. Because of the geometric constraints, olefin geometry is controlled (Eq. 99). Olefination of the bifunctional

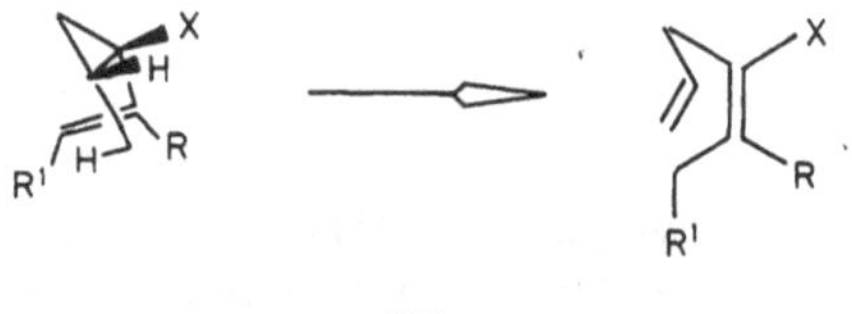
(99)

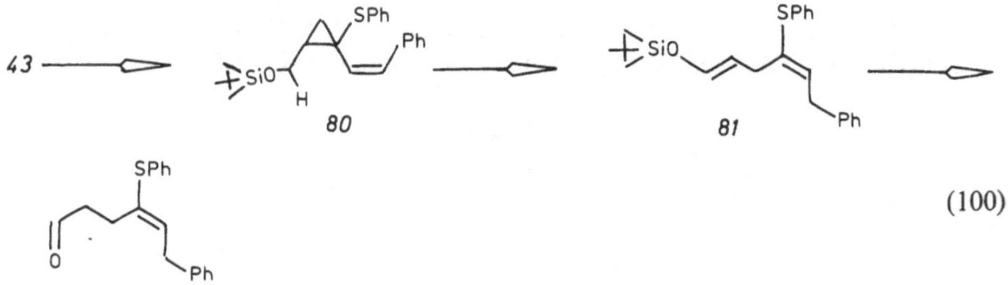
(100)

conjunctive reagent *43* creates exactly this structural feature and unravels to the diene *81* upon thermolysis of *80* (Eq. 100) [91]. Most notably is the fact that a single geometric isomer of the vinyl sulfide is produced. Equation 101 summarizes the

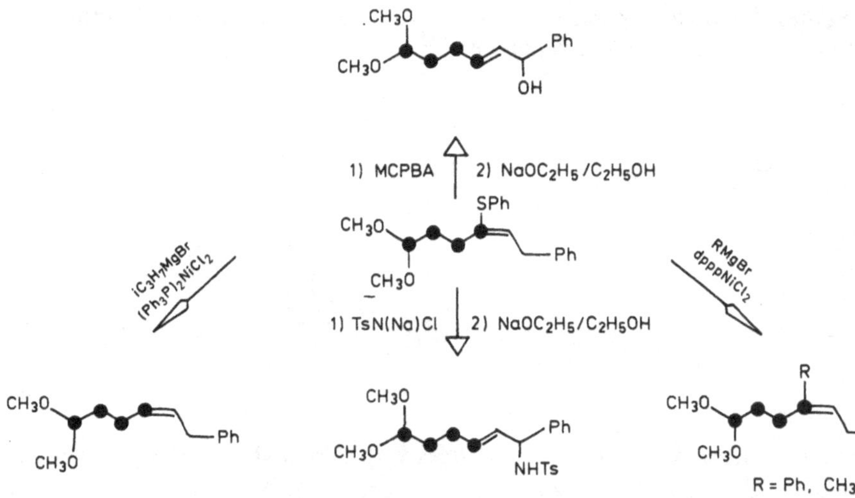

(101)

special uses of this type of product and the ability to translate the geometry of the vinyl sulfide into the geometry of di- and tri-substituted olefins lacking the sulfur. The four carbons of *43* translate into the carbons dotted in the structures of Eq. 101 in which the termini of the four carbons of *43* function as a 1,4-dication synthon *82*.

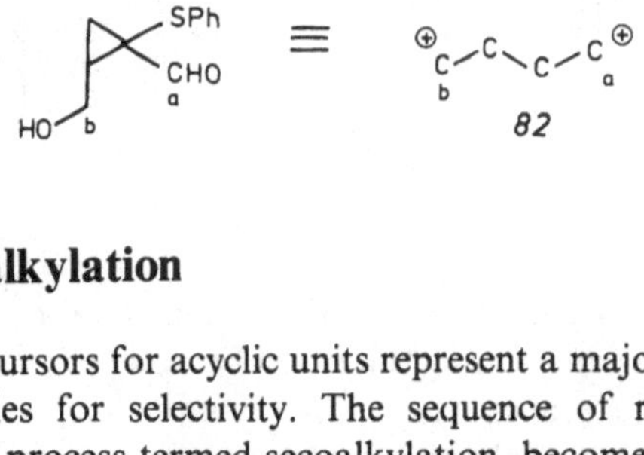

82

11 Secoalkylation

Cyclic precursors for acyclic units represent a major synthetic strategy because of the opportunities for selectivity. The sequence of ring formation followed by ring cleavage, a process termed secoalkylation, becomes a useful approach. The simplest way to take advantage of the three membered ring conjunctive reagents for such a process stems from the cyclobutene spiroannulation (Eqs. 41 and 42). The corresponding cyclobutene *83* from tetrahydroeucarvol smoothly undergoes conrotatory opening to create the diene *84* [74]. In this case, secoalkylation accomplishes an equivalent of a novel olefination. The cyclobutanones are particularly attractive

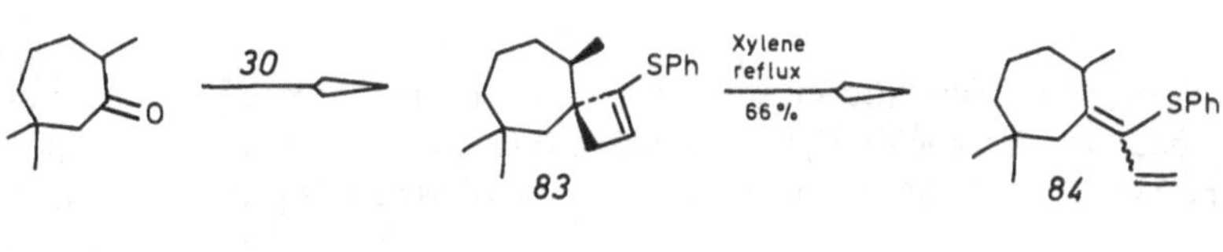

83 *84* (102)

intermediates for novel sculpting of the carbon framework by taking advantage of the activating influence of the carbonyl group to initiate relief of the high strain of the four membered ring. In a simple approach, Lewis acid coordination with the carbonyl oxygen initiates nucleophilically triggered ring opening (Eq. 103) [122,123]. At present only heteroatom nucleophiles have been reported. Equation 104 shows the application of this sequence to 2-heptanone [123].

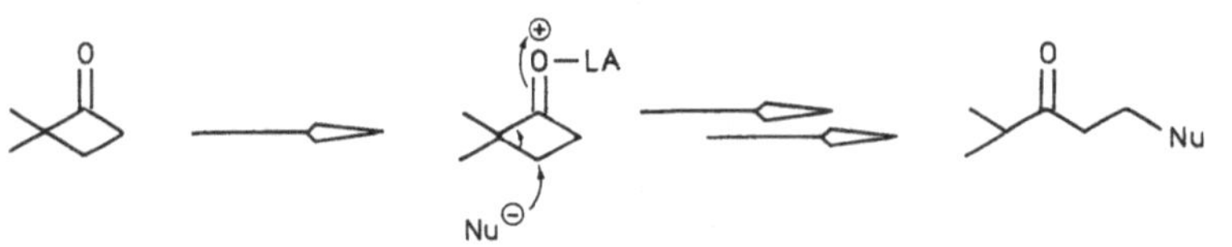

(103)

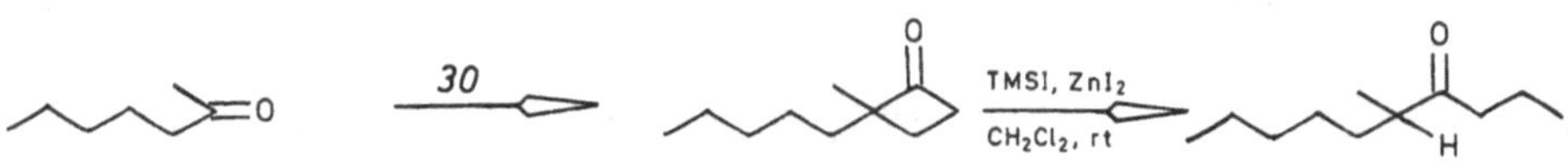

(104)

Treatment of the adducts of *30* and ketones with excess HCl in the presence of a Lewis acid leads directly to an acyclic system as in *85* (Eq. 105) [122]. It is likely that the thiophenol liberated in the ring expansion to the cyclobutanone *86* simply initiates the ring opening outlined in Eq. 103. Indeed, treating the cyclobutanone under the same conditions and adding thiophenol exactly mimics the direct transformation. Such β-phenylthioketones are easily manipulated as shown in Eq. 105. This process then serves as a reductive acylation of a carbonyl group as summarized in Eq. 106. The possibility that other nucleophiles can participate makes this an exciting avenue for future efforts.

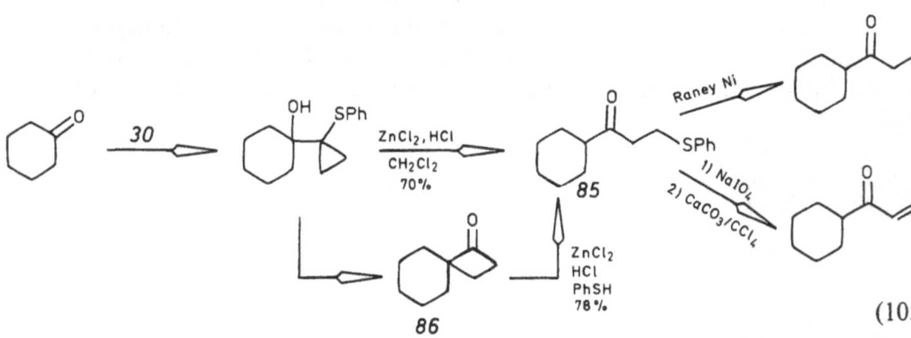

(105)

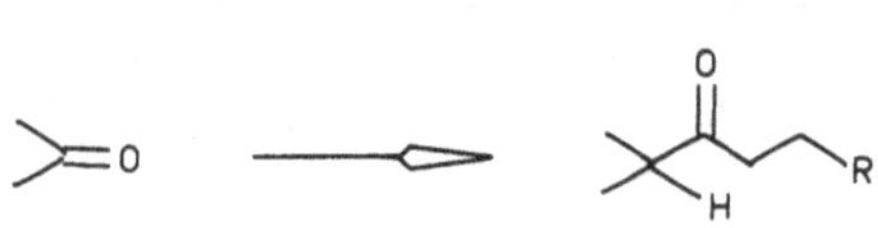

(106)

An alternative strategy for ring clevage envisions the presence of a potential leaving group beta to the cyclobutyl carbonyl group as in *87* (Eq. 107). Such systems may be synthesized directly from any carbonyl partner by the previously discussed substitutive spiroannulation of any simple carbonyl compound or spiroannulation of a carbonyl compound already possessing such a group. An example of the former approach makes use of the bromocyclobutanone *88* from cyclopentanone

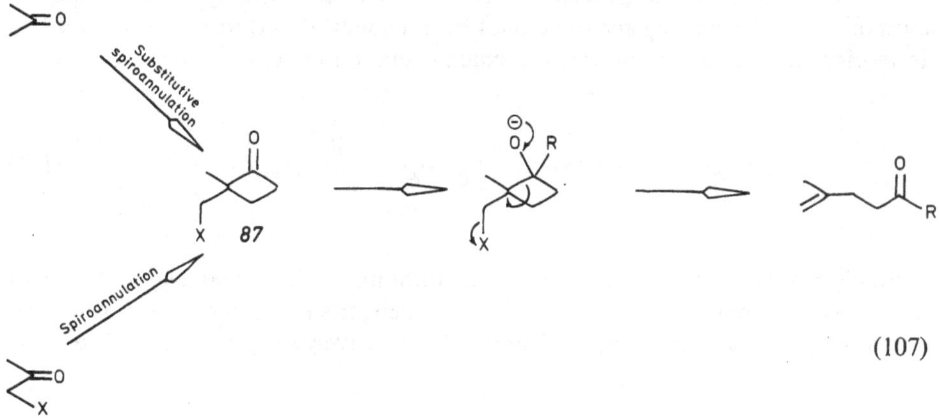

(107)

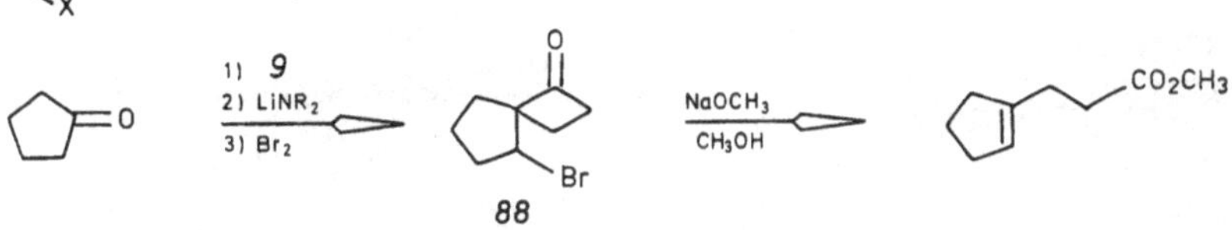

(108)

which fragments upon methanolysis as shown in Eq. 108 [42] α,β-Epoxyketones prove to be valuable spiroannulation substrates for secoalkylation. For example, spiroannulation of *89* directly creates the requisite functionality [36, 124]. Simply dissolving cyclobutanone *90* in methanolic methoxide at room temperature effects complete fragmentation in minutes — the facility being a reflection of the high relief of strain (Eq. 109).

(109)

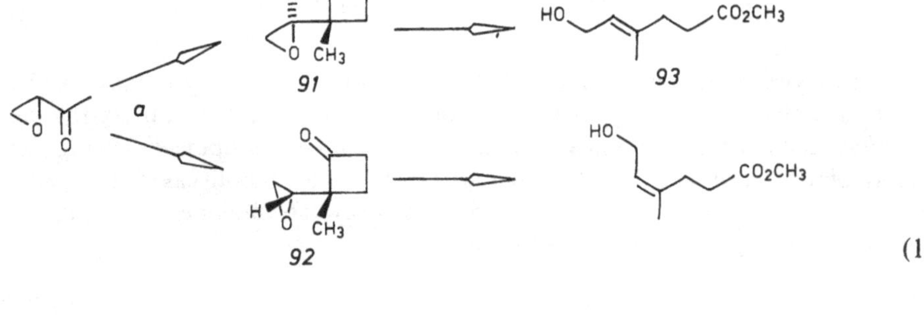

(110)

The fragmentation is stereospecifically anti as shown by complementary geometry obtained in the cleavage of the epimeric pair of epoxycyclobutanones *91* and *92* (Eq. 110). The fragmentation product *93* of cyclobutanone *91* is transformable into the dimethyl ester of the pheromone of the Monarch butterfly. Considering the availability of the starting epoxy ketones from enones, the oxasecoalkylation serves to reorient the oxidation pattern with chain extension as summarized in Eq. 111.

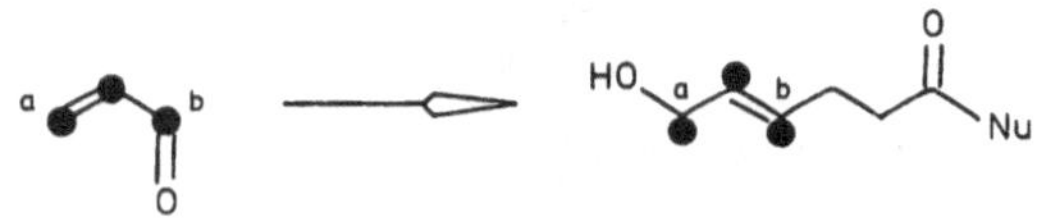

(111)

An alternative strategy to these same structural units takes advantage of the use of the electrophilic conjunctive reagent *37b* which converts ketones to α-vinylcyclobutanones such as *95* and *96* in Eqs. 112 and 113 respectively. Addition of a nucleophile

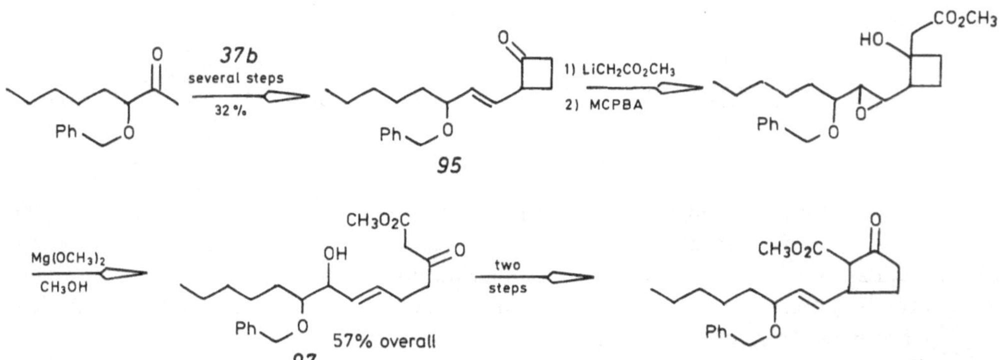

(112)

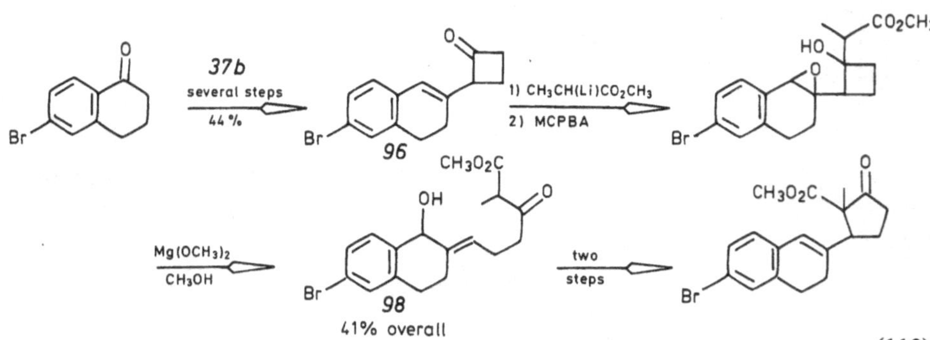

(113)

to the carbonyl groups is not hampered by enolization — probably because of the increased reactivity due to ring strain. Epoxidation completes the priming for the fragmentation to the acyclic units *97* and *98* which proceeds upon dissolving the epoxycyclobutanols into methanolic magnesium methoxide. In both cases, the acyclic units of *97* and *98* were reconstituted into rings — cyclopentanones — in pursuit of the skeletons of prostaglandins and steroids.

Replacing the epoxide by a cyclopropane, also available from an enone as in the preparation of *99* (Eq. 114), also possesses in excess of 209 kJ/mole (50 kcal/mole)

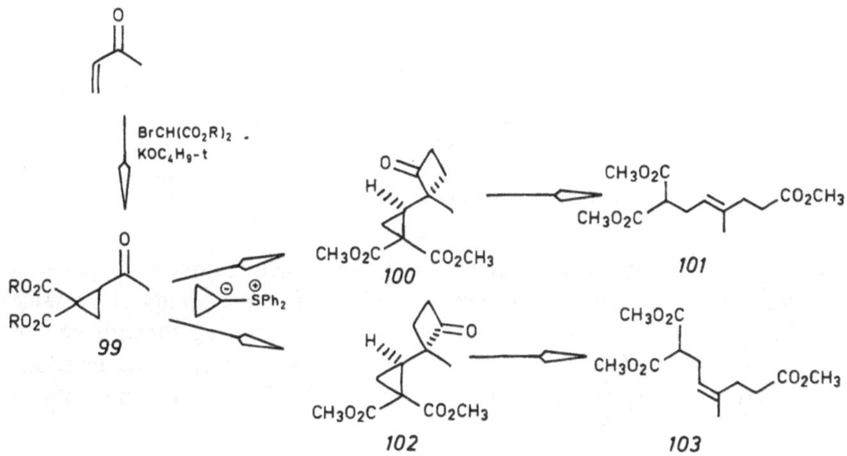

(114)

of strain release as a driving force for a fragmentation although a carbon leaving group, unprecedented in fragmentations, is mandated. Nevertheless, remarkably facile fragmentations of *100* and *102* occur in methanolic sodium methoxide in close parallel to the epoxide with only a change in temperature from 25 °C to 65 °C required [46]. The reaction is stereospecific — *100* gives cleanly the *E* olefin *101* and *102* only the *Z* olefin *103*. This carbasecoalkylation effects a double chain extension of an enone as summarized in Eq. 115a.

Alternatively, an anion stabilizing group alpha to the carbonyl group of the cyclobutanone provides a pathway for cleavage by attack of a nucleophile on the cyclobutanone carbonyl carbon. C-Acylation creates the familiar 1,3-dicarbonyl system which, by deacylation involving attack at the cyclobutanone carbonyl group, leads to a geminal alkylation as shown in Eq. 115b [47, 79].

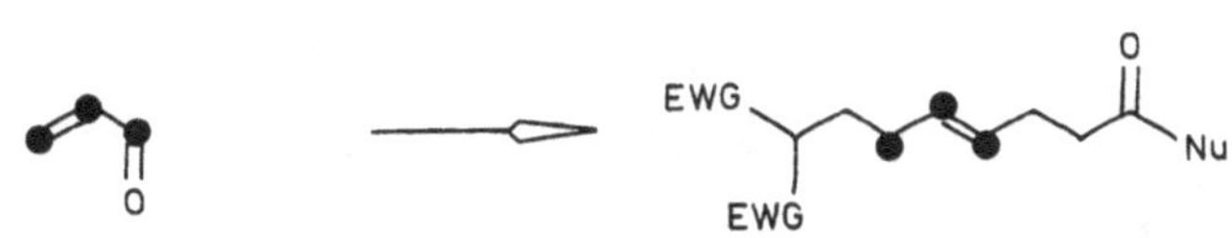

(115a)

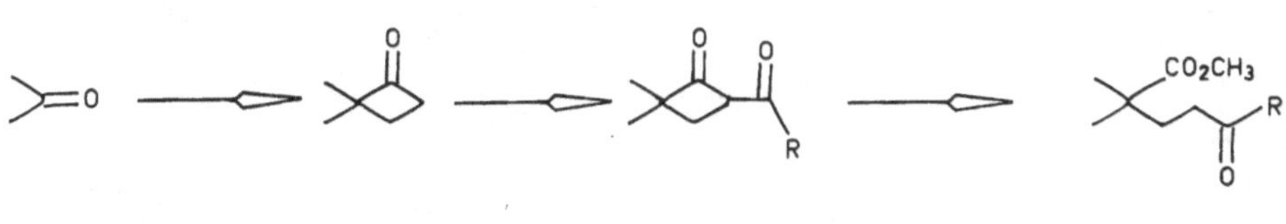

(115b)

While the haloform reaction normally only cleaves methyl ketones because of the structural requirements for the α,α,α-tribromomethyl ketone to induce fragmentation, the strain release that accompanies cleavage of a cyclobutanone permits extension of this concept to effect a net geminal alkylation as in Eq. 116[40, 125, 126]. The gem dibromo group offers flexibility for structural variation as summarized in Eq. 117.

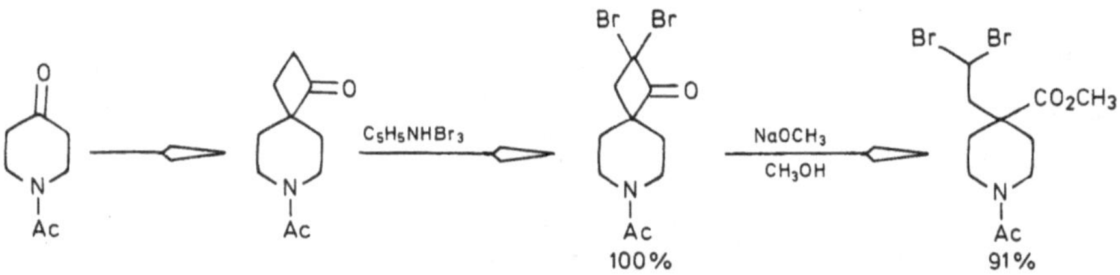

(116)

The fact that Favorskii-type ring contraction[127] may accompany the fragmentation of α,α-dibromocyclobutanones because of the leaving group ability of bromide suggests replacement of the bromine by sulfur, an anion stabilizing substituent that is a poor leaving group[128–130]. The cyclobutanone *104* derived from tetralone encounters just that problem in the haloform approach. The introduction of the

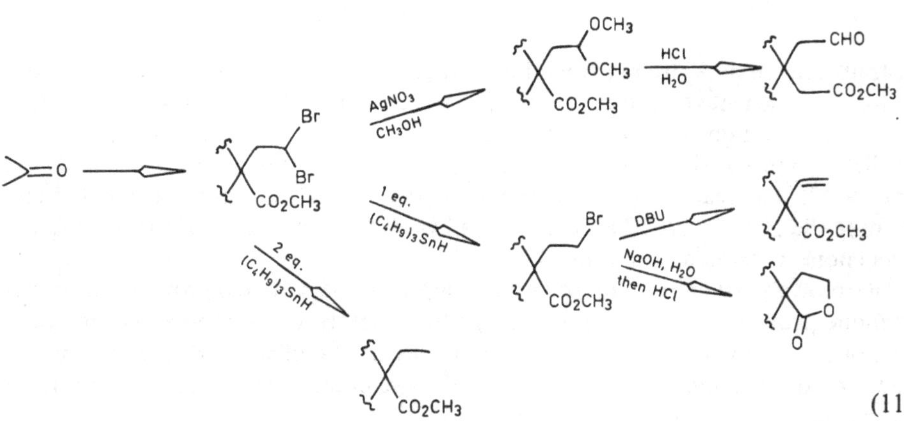

(117)

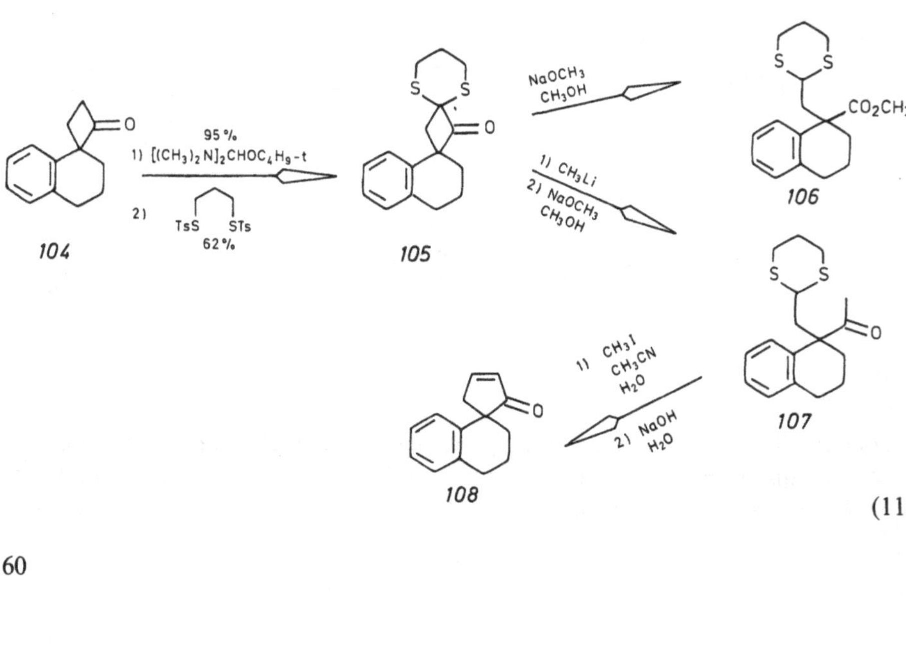

(118)

α,α-dithianyl group as in *105* begins by formylation with the Brederick reagent, bis(dimethylamino)-t-butoxymethane, followed by sulfenylation with trimethylenedithiotosylate (Eq. 118). While such α,α-bis-thioketones normally do not cleave with methoxide [131], the high strain of the cyclobutanone provides sufficient driving force to convert *105* to *106*. Initiation of fragmentation by use of carbon nucleophiles greatly expands the scope of this geminal alkylation as shown in the preparation of the methyl ketone *107* [129]. Unmasking the aldehyde of the differentiated chains of *107* allows an aldol condensation to complete a spiroannulation of a cyclopentenone onto a carbonyl group as in eq. 118 which summarizes the result of conversion of tetralone to *108*.

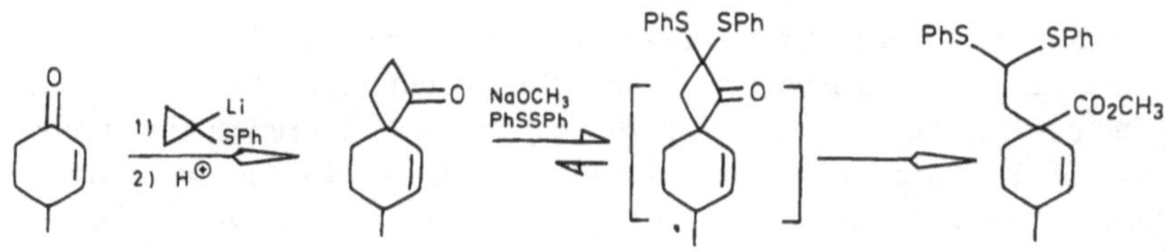

109 110 111 112

(119)

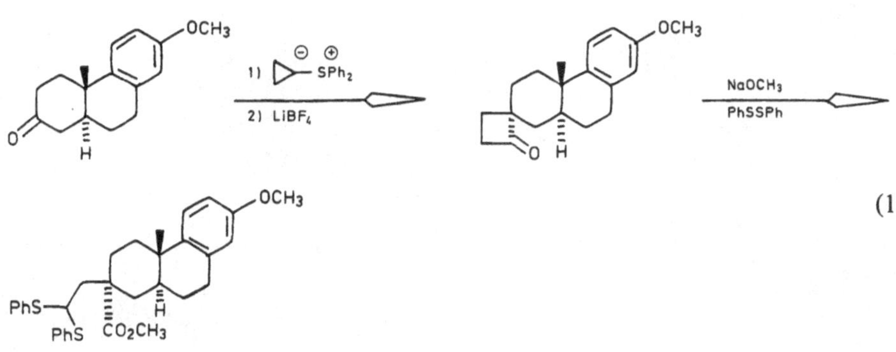

(120)

The ability to sulfenylate ketones reversibly using alkoxide bases in alcoholic solvents suggests telescoping the sulfenylation-fragmentation into a single pot operation [39]. The cyclobutanone *110* generates an equilibrium concentration of the bis-phenylthiolated ketone *111* which is susceptible to a nucleophilically triggered ring cleavage. The latter reaction drives the equilibrium to the right to give *112* directly from *110*. Thus, a two pot operation converts *109* into *112* in an 80% yield. The high stereocontrol of the cyclobutanone annulation converts the sequence of spiroannulation-secosulfenylation into a stereocontrolled geminal alkylation as shown in Eq. 120. Extraordinary chemoselectivity characterizes this sequence as demonstrated by the secosulfenylation of *113* in Eq. 121 [39b].

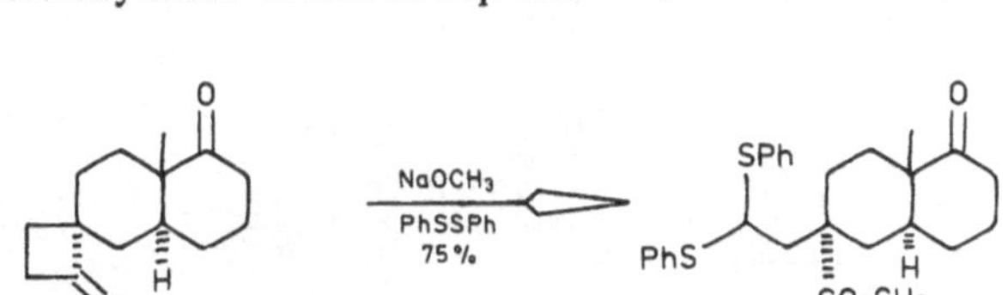

113

(121)

How much of an anion stabilizing group is required to provide a kinetic pathway for the release of strain of a cyclobutanone? The sulfenylated cyclobutanones offer a probe. Whereas, the bis-phenylthiocyclobutanone *111* cleaves under conditions of

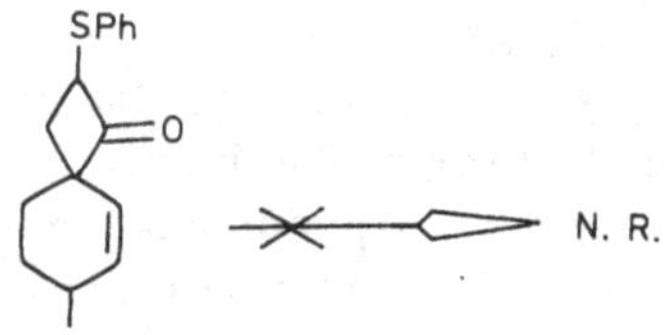

114

its birth, the monosulfenylated analogue *114* fails to cleave under any conditions. The enhanced stabilization of a carbanion by a sulfoxide compared to a sulfide, as revealed by increased carbon acidities of up to 13 pKa units [132)], is reflected in a rapid fragmentation of *115*, Eq. 122, under conditions similar to the bis-thiolated cyclobutanones [39 b)]. The versatility of this fragmented product in terms of further reactions such as a Pummerer rearrangement to *117* or an elimination to *118* suggests general utility for this activation as well.

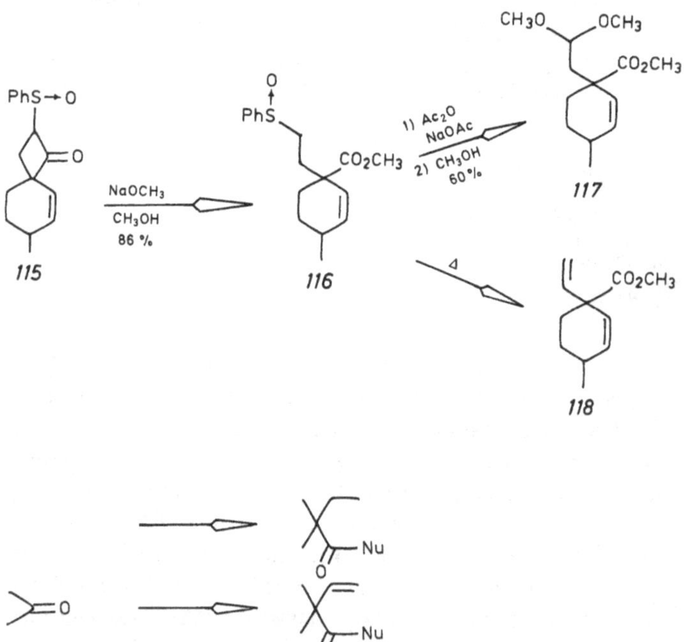

(122)

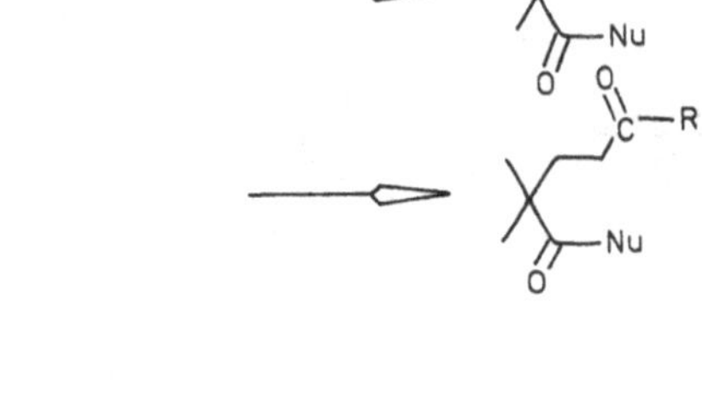

(123)

The sequences of spiroannulation-secobromination or, better, spiroannulation-secosulfenylation offer great versatility for geminal alkylation as summarized in Eq. 123. The fact that spiroannulation of carbonyl compounds proceeds with stereocontrol converts these transformations into stereocontrolled carbonyl group elaborations.

Some of the other cyclopropyl conjunctive reagents also provide ready access to fragmentation reactions as summarized in Eqs. 22 and 23. A brominated analogue of *6*, i.e. *119*, generates carbonyl adducts that can fragment by several modes [133]. For example, *120* solvolyzes in a polar solvent by ionization of the bromide with ring opening to give the allyl cation *121*. Solvent capture at the center of highest positive charge and acetylation of the unstable alcohol gives the open chain extended *122*. Simple synthetic manipulation to *123* establishes the validity of this approach directed towards juvenile hormones.

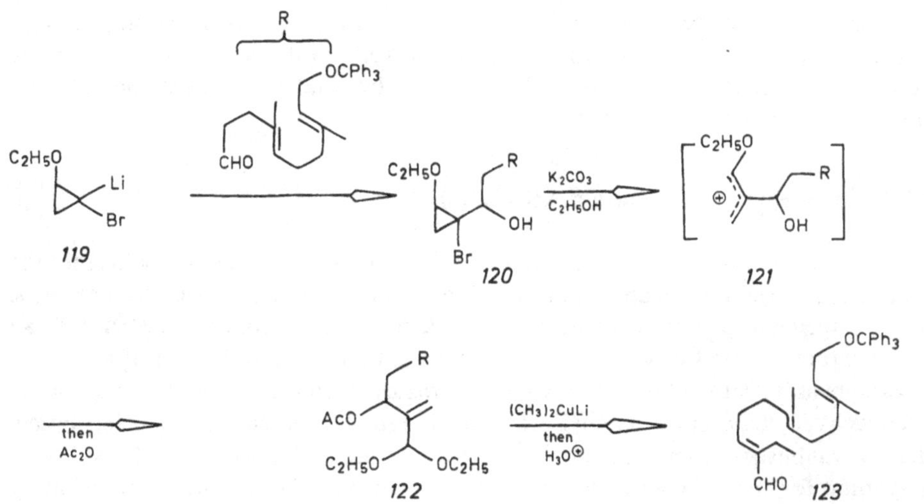

(124)

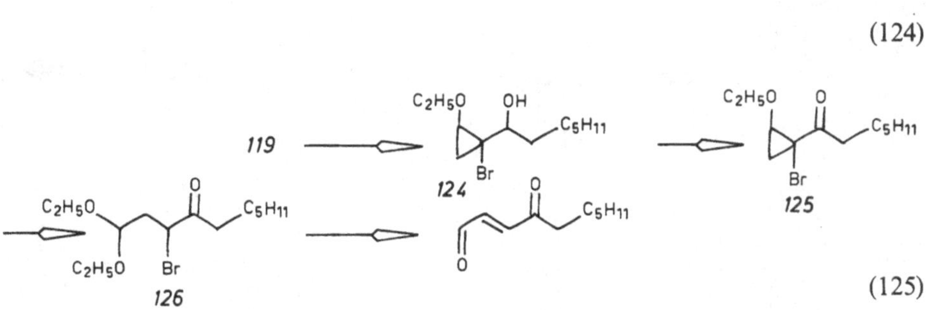

(125)

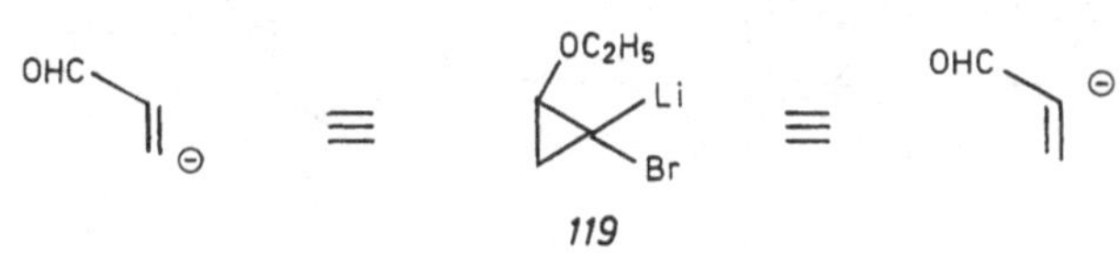

(126)

Alternatively, the alcohol group of the initial adducts of *119* such as *124* may be oxidized to the ketone *125* (Eq. 125). The resultant β-alkoxyketone now relieves ring strain by a retro-aldol type of reaction in the presence of acid catalysts to give an alternative type of chain extension. Thus, manipulating the same adducts allows *119* to serve as either an α- or β-acrolein anion synthon (Eq. 126).

The acyclic structural variations possible with various cyclopropyl conjunctive reagents make further developments in this methodology a rich fishing ground. The fact that the structural correlations between a synthetic target and the starting material are not as straighforward as classical methodology creates a need for rethinking synthetic strategy and thus retrosynthetic analysis.

12 Total Syntheses

The utility of small ring conjunctive reagents in total syntheses appears to be very great. The convulated structural correlations between the target and a starting material hamper its ready adaptation. The facility and stereochemical implications make the rethinking of retrosynthetic analysis well worth the effort. To illustrate some of this potential, several applications in the total synthesis of natural and unnatural products are analyzed.

12.1 Secoalkylation-Geminal Alkylation

Methyl deoxypodocarpate *127* (Scheme 1) [129] represents a simple problem since the ketone *132* is well-known and readily available from Hagemann's ester in three steps. The problem of geminal alkylation of this ketone stems from its existence as an *E*/*Z* mixture of ring fusion isomers. Recognizing that decarbonylation of aldehydes occurs readily with Wilkinson's catalyst creates a structural equivalence of an acetaldehyde chain and a methyl group as in *128*. This simple relationship immediately establishes several options, a simple one uses a thioacetal such as *129* as a synthon for the aldehyde. The presence of a carbonyl group three carbons away

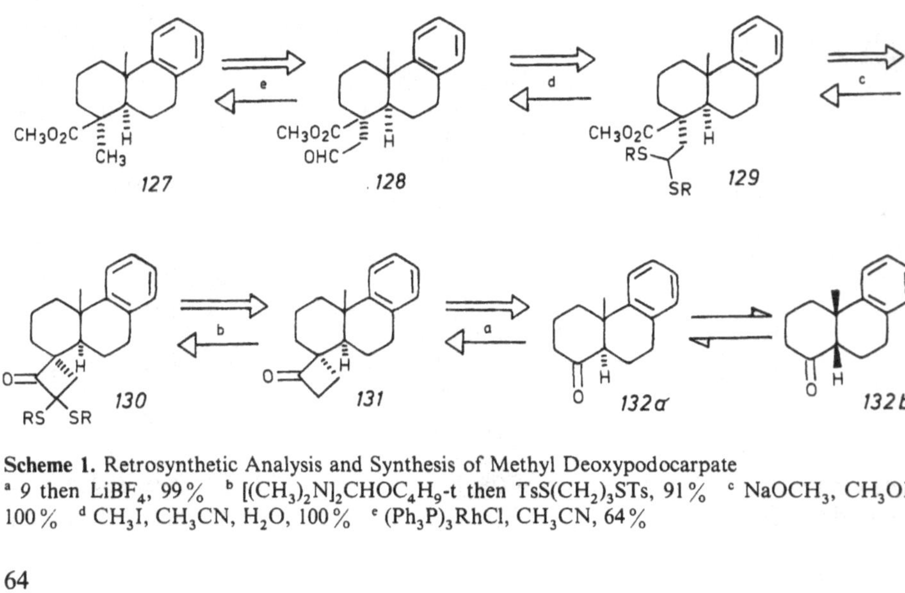

Scheme 1. Retrosynthetic Analysis and Synthesis of Methyl Deoxypodocarpate
[a] *9* then $LiBF_4$, 99% [b] $[(CH_3)_2N]_2CHOC_4H_9$-t then $TsS(CH_2)_3STs$, 91% [c] $NaOCH_3$, CH_3OH, 100% [d] CH_3I, CH_3CN, H_2O, 100% [e] $(Ph_3P)_3RhCl$, CH_3CN, 64%

from the acetal suggests a cleavage of an α,α-dithiocyclobutanone *130* as the immediate precursor of *129* and, obviously, the parent cyclobutanone *131* as the ultimate precursor. The attractiveness of such an analysis stems from the stereochemical consequences of the cyclopropyl sulfur ylide additions. Even though *132* exists as an *E*, *Z* mixture, equilibration of the ring junction occurs rapidly compared to carbonyl addition of *9*. Since the *E* isomer reacts more rapidly than the *Z* one and since high diastereoselectivity always accompanies the carbonyl additions of *9* with the carbonyl group of the cyclobutanone introduced on the more hindered face of the starting ketone, a single cyclobutanone *131* results quantitatively, in which all the stereochemistry is fixed. While the secosulfenylation was accomplished by the stepwise sequence, the one step formation of *129* (R = Ph) would be a preferred approach. The remaining steps are unexceptional with the poorest yield being the decarbonylation. The overall yield of methyl deoxypodocarpate from Stork's ketone with complete stereocontrol is 58%.

The same concept applies to the synthesis of an intermediate *133* that stands two steps away from hinesol (Scheme 2) [129]. The recognition that an ester β to a carbonyl

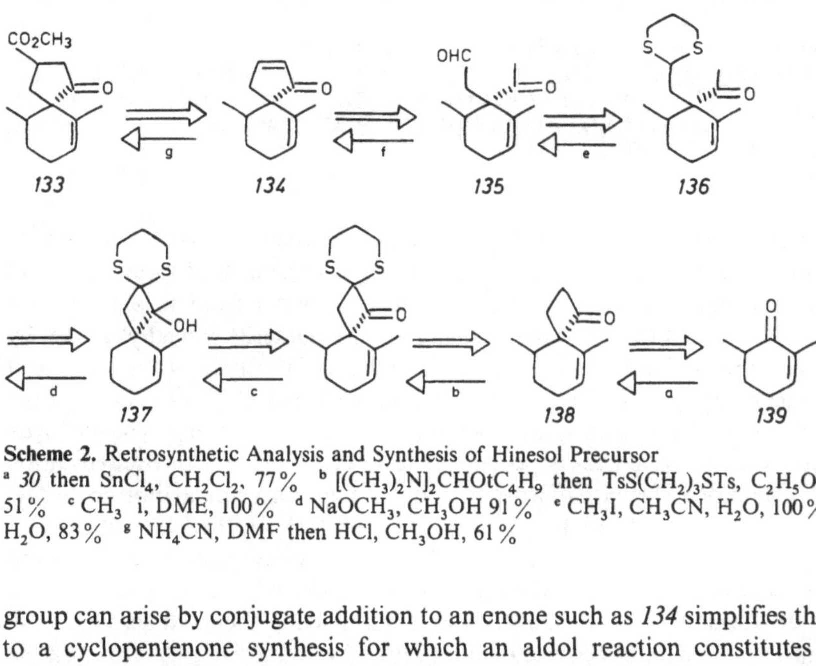

Scheme 2. Retrosynthetic Analysis and Synthesis of Hinesol Precursor
[a] *30* then $SnCl_4$, CH_2Cl_2, 77% [b] $[(CH_3)_2N]_2CHOtC_4H_9$ then $TsS(CH_2)_3STs$, C_2H_5OH, NaOAc, 51% [c] CH_3 i, DME, 100% [d] $NaOCH_3$, CH_3OH 91% [e] CH_3I, CH_3CN, H_2O, 100% [f] NaOH, H_2O, 83% [g] NH_4CN, DMF then HCl, CH_3OH, 61%

group can arise by conjugate addition to an enone such as *134* simplifies the problem to a cyclopentenone synthesis for which an aldol reaction constitutes the most direct approach. With the ketoaldehyde *135* identified as the target and the equivalency of a thioacetal with an aldehyde, *136* becomes the goal. Using the concept of secoalkylation, the cyclobutanol *137* and the parent cyclobutanone *138* represent the logical precursors. Spiroannulation of the enone *139* with proper stereochemistry (i.e. the carbonyl group of the cyclobutanone on the less hindered face) derives from use of lithiated cyclopropyl phenyl sulfide. Indeed, the direct reverse of this analysis creates *133* in an overall yield of 18%.

The haloform analogue of the above cleavage may also be employed to solve such problems of geminal alkylation. Scheme 3 outlines this approach to grandisol (*140*), a constituent of the sex pheromone of the boll weevil [74, 134]. To minimize problems

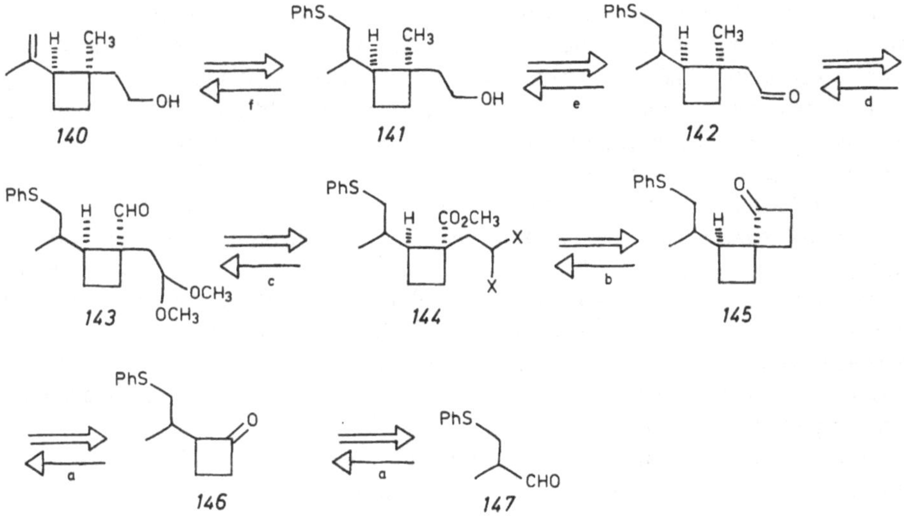

Scheme 3. Petrosynthetic Analysis and Synthesis of Grandisol
[a] *30* then TsOH, PhH, H_2O, 90, 92% [b] $C_5H_5NHBr_3$, HOAc then $NaOCH_3$, CH_3OH followed directly by $AgNO_3$, CH_3OH, 86% [c] LAH ether then SO_3, C_5H_5N, DMSO, $(C_2H_5)_3N$, 92% [d] NH_2NH_2, KOH, $HO(CH_2CH_2O)_3H$ then HCl, H_2O, THF, 83% [e] LAH, ether, 100% [f] MCPBA, CH_2Cl_2 then, $CaCO_3$, $PhCH_3$, $PhCH_3$, 84%

associated with the double bond, the ability to employ sulfoxide eliminations to create olefins permits *141* to become a logical precursor. The gem disubstitution of *141* becomes our concern. The two carbon chain can easily derive from an acetaldehyde side chain as in *142* and the one carbon chain, i.e. the quaternary methyl group, by a Wolff-Kishner reduction leading to the differentiated dialdehyde *143* as the logical precursor. The ready interconversion of oxidation levels permits replacement of the free aldehyde of *143* by its corresponding ester as in *144* ($X=OCH_3$). The aldehyde functionality of *144* masked as its acetal can arise either from a secosulfenylation sequence or a haloform cleavage of a cyclobutanone since the geminal dibromide *144*, X=Br, can solvolyze easily to the acetal *144*, $X=OCH_3$. Either way, the spirocyclobutanone *145* becomes the precursor. The spiroannulation substrate to lead to *145* is itself a cyclobutanone *146* which then can also derive from a cyclobutanone annulation of the aldehyde *147*, an adduct of thiophenol and methacrolein. The stereochemistry desired in the conversion of *146* to *145* with the carbonyl group on the least hindered face indicates the use of *30* as the spiroannulating conjunctive reagent. As demonstrated in Scheme 3, this analysis carried out in the synthetic direction led to grandisol in 46% overall yield!

The novel diterpene trihydroxydecipiadiene *148* consists of an unusual carbon framework. Scheme 4 [79] focuses on the geminal carbon substituents since conventional methods to elaborate this part of the molecule failed. The side chain allylic alcohol can be easily recognized to derive from olefination methodology utilizing stabilized ylides to create the *E* geometry. The aldehyde that results would exist as its lactol as in *149*. Many approaches to the remaining allyl alcohol of *149* exist starting from a carbonyl group as in *150* such as olefination, epoxidation, and base catalyzed

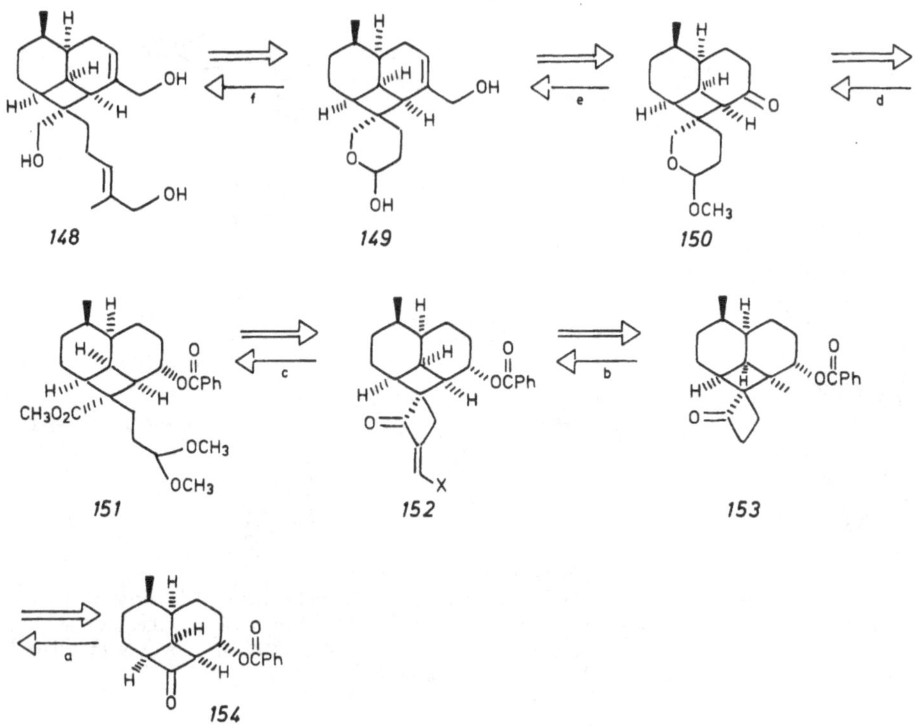

Scheme 4. Retrosynthetic Analysis and Synthesis of Trihydroxydecipiadiene
[a] *30* then H^+, 52%; [b] $[(CH_3)_2N]_2CHO\text{-t-}C_4H_9$, 100%; [c] CH_3OH, TsOH, 92%; [d] DIBAL-H ether then CH_3OH, Dowex-50W-X-8-H^+ then PCC 87%; [e] See ref. 79; [f] HCl, H_2O, THF then $Ph_3PC(CH_3)CO_2CH_3$, $PhCH_3$, 100° then DIBAL-H, ether 56%

opening of the epoxide. Masking this carbonyl group of *150* in the form of its less reactive congenor, the alcohol, allows us to focus on the two arms of the geminal carbon. Modifying the internal acetal to its open chain analogue and oxidizing the primary alcohol to an ester as in *151* immediately opens the opportunity of creating *151* by a retro-Clasen condensation from an α-formylcyclobutanone as in *152* (X=OH). The derivation of *152* by a cyclobutanone spiroannulation onto *154* now becomes obvious and the only question to resolve is choice of method to create the correct stereochemistry depicted in *153*. Having the carbonyl group of the cyclobutanone on the less hindered face indicates the use of *30*. In realizing this scheme, formylation of *153* was best accomplished using the Brederick reagent to give *152* $X=N(CH_3)_2$. Under the conditions of the retro-Claisen condensation, the vinylogous amide hydrolyzes in situ to the formyl compound (*152*, X=OH) and then cleaves. The spiroannulation approach creates *150* in 42% overall yield from the ketone *154*; whereas, other methods totally failed.

This same concept applies to spiroannulations of six membered rings and is illustrated in a synthesis of acorenone B *155* as outlined in Scheme 5 [47]. The notion of alkylative carbonyl transposition permits the spiro enone *156* to become a logical intermediate. The standard analysis by a retro-aldol process translates the spiro ring system of acorenone B into a geminal alkylation problem as revealed by

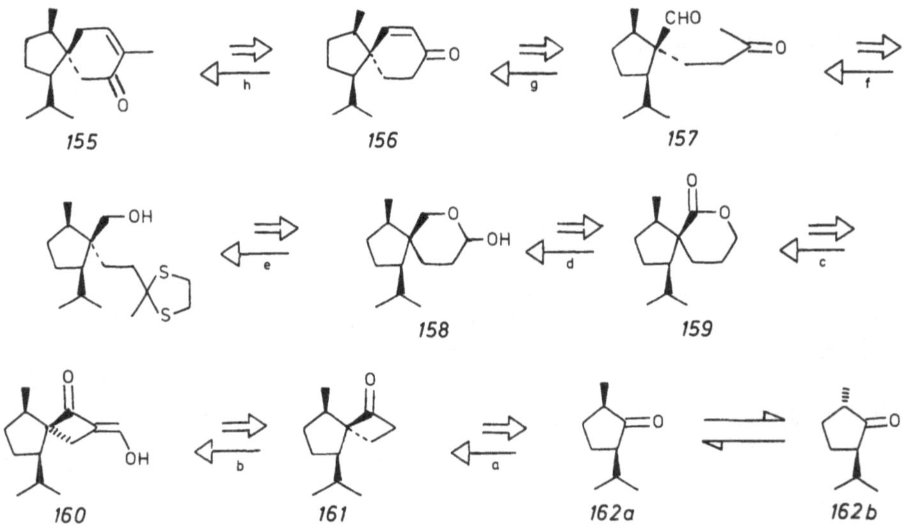

Scheme 4. Retrosynthetic Analysis and Synthesis of Acorenone B
[a] *9* then $LiBF_4$, 88%; [b] NaH, $HCO_2C_2H_5$, PhH; [c] TsOH, PhH, H_2O, 62% for steps [b] and [c]; [d] DIBAL-H, $PhCH_3$; [e] CrO_3, H_2O, H_2SO_4, CH_3COCH_3 then CH_3Li, ether followed by $HSCH_2CH_2SH$, BF_3, ether, 56% for steps [d] and [e]; [f] $C_5H_5NSO_3$, DMSO, $(C_2H_5)_3N$ then $HgCl_2$, CH_3CN, H_2O, 100%; [g] KOH, CH_3OH, 92%; [h] see ref. [47].

intermediate *157*. Keeping in mind the creation of a methyl ketone by carbonyl addition to a lactone and the normal synthons for carbonyl groups using various oxidation levels suggests the hydroxy aldehyde *158*, which would exist as its lactol and the enol lactone *159* as reasonable intermediates. The enol lactone is an equivalent of *151* in the trihydroxydecipiadiene synthesis and, as there, suggests a formylated cyclobutanone such as *160* as its precursor and the cyclobutanone *161* as the key. The question becomes one of stereocontrol since the spiroannulation substrate *162* is difficult to obtain as a single isomer and, even if one stereoisomer is available, readily epimerizes. Once again, the advantages of this spiroannulation become apparent. Under conditions of condensation of *162* with the cyclopropylide this cis isomer *162a*, which exposes a face unencumbered by any alkyl substituents, selectively reacts. The ylide methodology places the carbonyl group of the resultant cyclobutanone on the more hindered face. Thus, in a single operation, three stereogenic centers are established with a single relative stereochemistry. The remaining steps proceed uneventfully to give *156* in 28% overall yield.

13 Lactone Annulation and Substitutive Spiroannulation

The facile conversion of carbonyl groups into lactones via cyclobutanones offers many opportunities for synthetic applications considering the importance of butanolides in natural products synthesis. The iridoids vividly illustrate this potential. Allamandin (*163*) [135c)] and its dehydrated relative plumericin (*164*) [135d)], compounds possessing antifungal, antibacterial, and antitumor activity, pack a great deal of

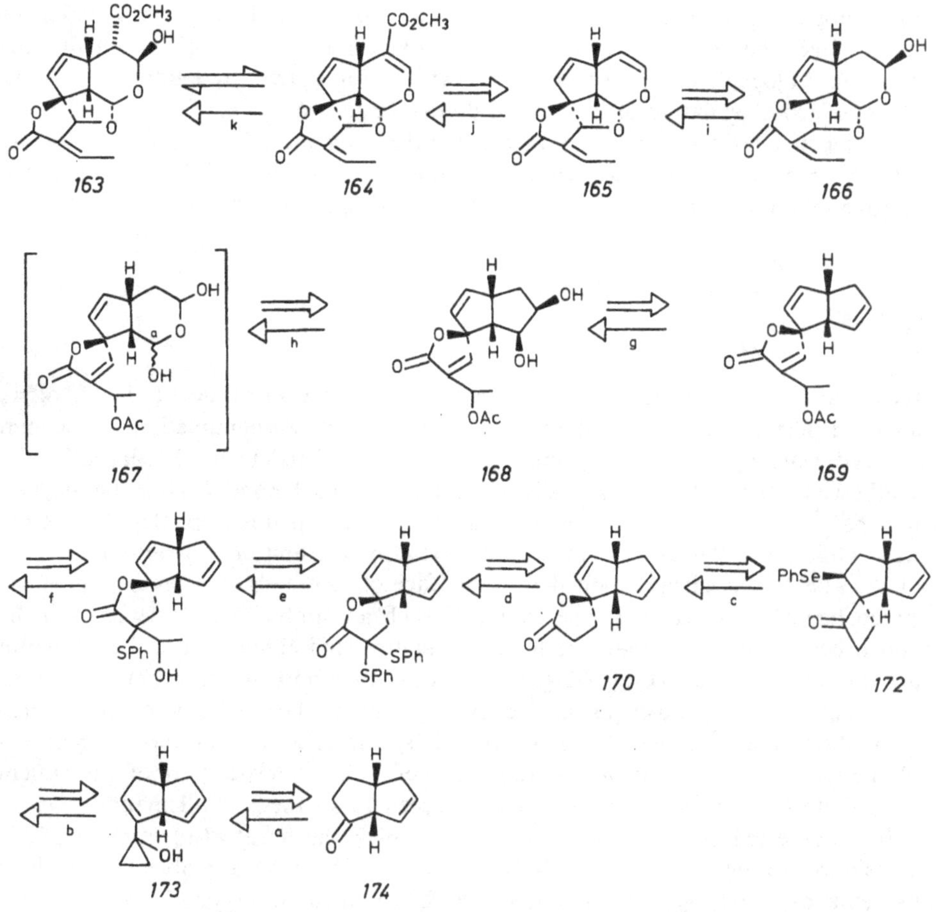

Scheme 5. Retrosynthetic Analysis and Synthesis of Plumericin, Allamandin, and Allamcin
[a] *9* then $LiN(C_2H_5)_2$, 82%; [b] PhSeBr, $(C_2H_5)_3N$, CH_2Cl_2, 80%; [c] MCPBA, CH_2Cl_2 65%; [d] LDA, $PhSSO_2Ph$, THF, 95%; [e] C_2H_5MgBr, ether, THF then CH_3CHO, 100%; [f] MCPBA, CH_2Cl_2 then, $CaCO_3$, CCl_4, followed by Ac_2O, DMAP, C_5H_5N, 75%; [g] OsO_4, NMO, THF, 90%; [h] $NaIO_4$, ether, H_2O then add NaOAc, 89%; [i] Ac_2O, DMAP, $(iC_3H_7)_2NC_2H_5$ then distill through hot tube (500°), 99%; [j] CCl_3COCl, 2,6(t-$C_4H_9)_2C_5H_3N$ then $Mg(OCH_3)_2$, CH_3OH, 77%; [k] $HClO_4$, H_2O, 57%

functionality in close proximity. Scheme 5 [43,135] outlines an analysis invoking lactone annulation combined with substitutive spiroannulation to generate a remarkably efficient approach where not a single protecting or activating group is employed.

Minimizing functionality by removal of the carbomethoxy group to descarbomethoxy plumericin *165* does greatly simplify the task but requires the development of a selective carbomethoxylation procedure for enol ethers. The enol ether and its hydrated form *166*, a co-occurring natural product known as allamcin [135e], can be seen to derive from a hydrated derivative of a bis-aldehyde *167*. If the hydroxyl group at C(a) of *167* equilibrates to the α-configuration, its juxtaposition with respect to the butenolide creates a possibility for conjugate addition-elimination to form the ring system of plumericin. The existence of plumeride suggests this route may have biogenetic implications. In an ideal case, *167* would directly cyclize upon its birth. Since an α,ω-dialdehyde arises by oxidative cleavage of an olefin, the bicyclo[3.3.0]octadiene *169* becomes the target. For steric and electronic reasons, chemoselective oxidation of the desired double bond is anticipated.

At this time, we can focus on the creation of the butenolide. Sulfur based methodoly allows simplification to the simple butanolide as in *170*. While cyclobutanone spiroannulation of *171* followed by a Baeyer-Villiger reaction would create *170*,

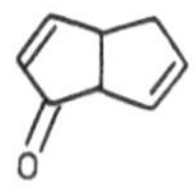

171

the replacement of one of the double bonds by a selenium substituent as in *172* offers a more efficient solution using the concept of substitutive spiroannulation. A selenium induced rearrangement of the vinylcyclopropanol *173* to create *172* simplifies the synthesis to the well-known, easily available bicyclic ketone *174* as the starting meterial. The stereochemistry of the synthesis resides in this step. Based on a net trans addition to the trisubstituted double bond of *173* and on a preferential attack of the selenium electrophile on the convex face of *173* leads to a prediction of the stereochemistry as depicted. The choice of such an approach also stems from the added benefit that the conditions for the oxidation and elimination of the selenium are the same for the Baeyer-Villiger reaction — thus both structural modifications can occur simultaneously as a one pot operation. The validity of this spiro-annulation based strategy is demonstrated by the efficiency of the synthesis — plumericin in 13 steps in an overall yield of 17% [43]. Hydration of plumericin with dilute aqueous perchloric acid also completes a synthesis of allamandin [135].

In a projected synthesis of verrucarol in which the tetrahydropyranone *175* is envisioned to play a role, the lactone *176* can serve as a viable intermediate (Scheme 6) [75]. Using cyclobutanone annulation methodology, *177* becomes the required intermediate. The geminal substitution of *177* can also be recognized to derive from a spiroannulation approach as previously analyzed.

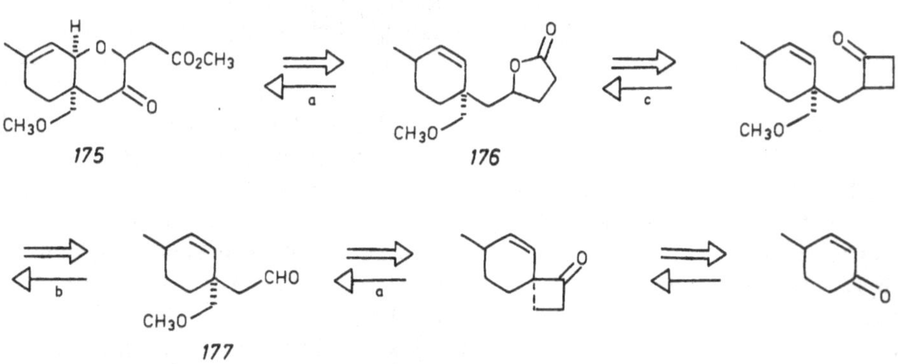

Scheme 6. Retrosynthetic Analysis and Synthesis of Potential Verrucarol Intermediate
[a] See secosulfenylation section and ref. [75]; [b] *30* then TsOH, PhH, H_2O, 62%; [c] NaOH, H_2O_2

The availability of cyclopentenones from butanolides allows the lactone annulation to facilitate the synthesis of cyclopentyl natural and unnatural products. An example that highlights the latter is dodecahedrane (*178*) for which *179* constitutes a critical synthetic intermediate [136, 137]. Lateral fusion of cyclopentenones as present in *179* can arise by acid induced reorganization and dehydration of *180*. While a variety of routes can be envisioned to convert a ketone such as *182* into *180*, none worked satisfactorily [137]! On the other hand, the cyclobutanone spiro-annulation approach via *181* proceeds in 64% overall yield. Thus, the total carbon cource of dodecahedrane derives from two building blocks — cyclopentadiene and the cyclopropyl sulfonium ylide.

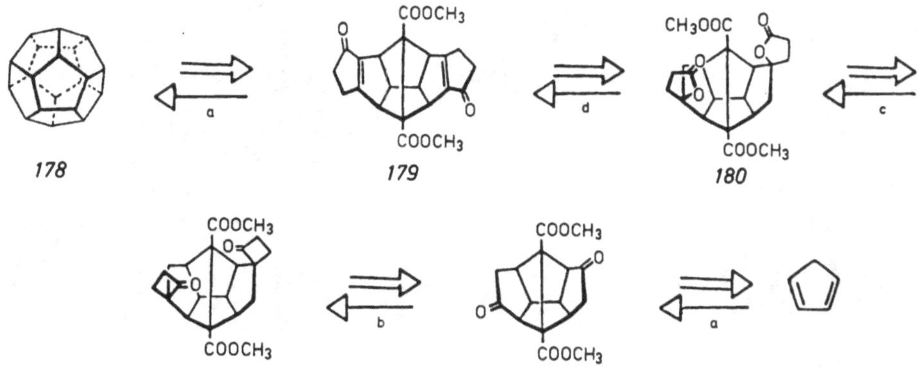

Scheme 7. Retrosynthetic Analysis and Synthesis of Dodecahedrane
[a] See Ref. [136]; [b] *9* then HBF_4, 77%; [c] H_2O_2, NaOH, CH_3OH, H_2O, 100%; [d] P_2O_5, CH_3SO_3H, 83%

14 Cyclopentyl Syntheses

The construction of cyclopentanones can take several avenues based upon the concepts evolving from use of small ring conjunctive reagents. A very simple one is α-cuparenone (*183*) which can be thought to derive by a ring expansion of a cyclobutyl carbinyl system as in *184* (Scheme 8) [110, 111]. If X is an anion stabilizing

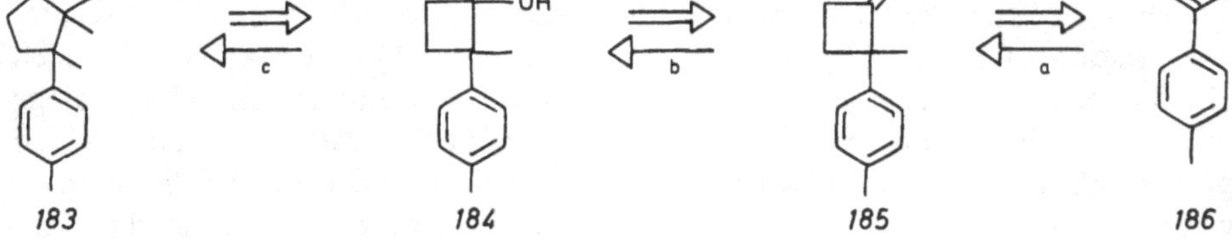

Scheme 8. Retrosynthetic Analysis and Synthesis of α-Cuparenone
[a] *30* or *33* then TsOH, PhH, H_2O, 44–68%; [b] *185* $(CH_3)_2C(Li)SeCH_3$, ether, 66%; [c] CH_3OSO_2F, ether, 82%

group as well as a potential leaving group as in the case of $X=SeCH_3$, the cyclobutanone *185* and thus p-methylacetophenone (*186*) become the precursors for this extraordinarily short approach [111]. This four step sequence telescopes into only three by using the anion of isopropylphenylselenoxide in which the intermediate *184* undergoes the pinacol type of rearrangement directly [110].

The vinylcyclopropane rearrangement is particularly valuable for cyclopentane synthesis. Aphidicolin (*187*), a most unusual antitumor agent, yields nicely to an analysis based on the concept of cyclopentanone annulation as outlined in Scheme 9

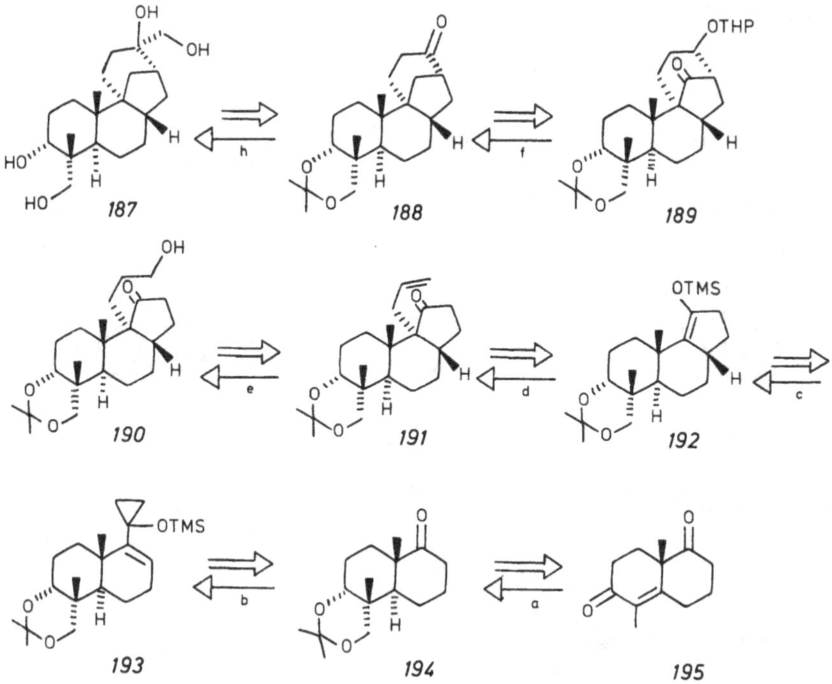

Scheme 9. Retrosynthetic Analysis and Synthesis of Aphidicolin
[a] See ref. [138]; [b] *9* then PhSeNa, DME followed by BSA, $(C_2H_5)_3N$, PhH, 56%; [c] FVT then $Pd(OAc)_2$, CH_3CN followed by Li, NH_3, t-C_4H_9OH, TMS-Cl, 58%; [d] C_4H_9Li, THF, HMPA, $CH_2{=}CHCH_2I$, 35%; [e] $(CH_3)_2CHC(CH_3)_2BH_2$, diglyme, 0° then add NaOH, H_2O_2, 57%; [f] PCC, NaOAc, CH_2Cl_2 then KOH, CH_3OH followed by DHP, TsOH, 54%; [g] NH_2NH_2, KOH, diglyme then TsOH, CH_3COCH_3, H_2O followed by PCC, NaOAc, CH_2Cl_2, 62%

[65]. During the structure determination of aphidicolin, it was reconstituted from the ketone *188* which becomes the principal target. A greatly revealing structural modification of the latter is the addition of a carbonyl group as in *189* which discloses the ability to construct the bicyclo[3.2.1]octyl unit via an aldol strategy utilizing an aldehyde derived by oxidation of the alcohol *190*. The availability of primary alcohols by anti-Markovnikov hydration of an olefin reveals the presence of an allylated cyclopentenone *191*. Introduction of the allyl group at the more substituted side of the cyclopentanone requires creation of the more substituted enolate which may arise from the corresponding enol silyl ether *192*, the exact type of skeleton created by rearrangement of a vinylcyclopropanol trimethylsilyl ether *193*.

The source of the latter from the appropriate ketone *194* using cyclopropyl sulfur ylide chemistry is apparent. The ultimate starting material is the well known diketone *195* which is easily available in optically active form.

Realization of this scheme initially encountered two obstacles. Opening of the oxaspiropentane *196* failed with base. The previously discussed merged substitution-

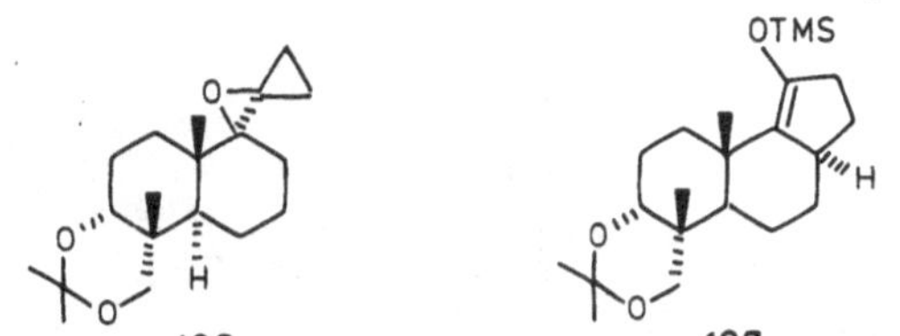

196 *197*

elimination approach was created to resolve this impasse and should prove to be generally useful. The rearrangement of *193* produced a mixture of *192* and *197* with the latter dominating — a most startling result since *197* requires the central ring to be in a boat conformation. This slight defect was easily corrected by oxidation to an enone followed by dissolving metal reduction to create the thermodynamically more stable *192*. The remaining steps proceeded uneventfully to provide *188* in 2% overall yield from *194*.

The electrophilic cyclopropyl conjunctive reagents offer different types of building blocks and therefore strategies for cyclopentane synthesis via vinylcyclopropane rearrangements. The spirovetivane *198*, a constituent of Vetiver oil and a versatile intermediate towards further derivatives, represents a target focussing on the α-alkoxycyclopropanecarboxaldehyde *39* as the key reagent (Scheme 10) [96]. The ready

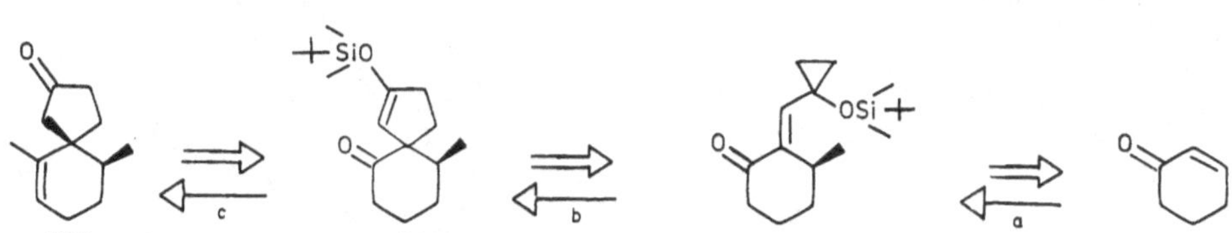

198 *199* *200*

Scheme 10. Retrosynthetic Analysis and Synthesis of a Spirovectivane
[a] CH_3MgI, CuI then *39b* followed by Ac_2O, DBN, 73%; [b] LDA, TMSCl, ether then FVT followed by CH_3OH, $(C_2H_5)_3N$, unstated yield; [c] CH_3MgI, ether then TsOH, PhH, 90%

availability of a methyl substituted cyclohexene by dehydration of a tertiary alcohol combined with the need to create a cyclopentene from the cyclopentanone leads to the keto enol silyl ether *199*. Simplification to the vinyl cyclopropane *200* becomes obvious. The presence of the α,β-unsaturated ketone in *200* suggests a cross-aldol coupling using a cyclopropyl carboxaldehyde. The problem of regioselectivity of the aldol condensation with respect to 3-methylcyclohexanone can be skirted by trapping a kinetic enolate derived from conjugate addition to cyclohexenone, the starting material. Execution of this synthesis proceeds as projected in a short approach to this important class.

An alternative approach to the spirovetivanes with electrophilic conjunctive cyclopropyl reagents is illustrated in a synthesis of α-vetispirene, *201* (Scheme 11) [139].

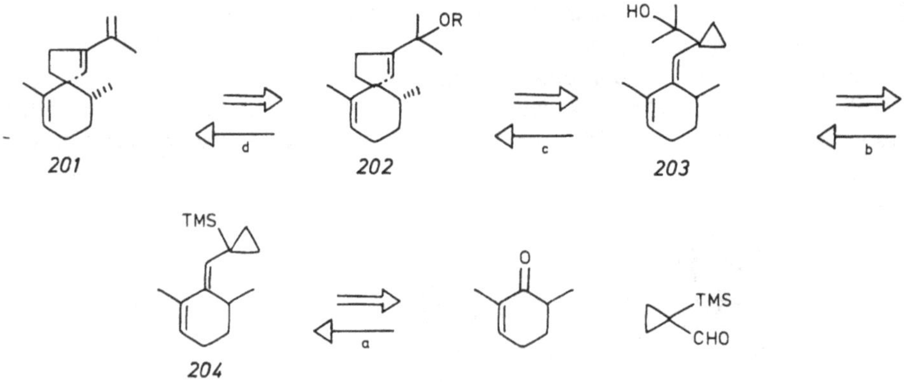

Scheme 11. Retrosynthetic Analysis and Synthesis of α-Vetispirene
[a] LAH, $TiCl_3$, THF, 0°, 50–60%; [b] $(nC_4H_9)_4NF$, $(CH_3)_2CO$, THF, 90%; [c] C_4H_9Li, THF, HMPA, CH_3I then 435–440 °C; [d] TsOH

The isopropenyl side chain may derive by elimination of a tertiary alcohol or ether as in *202*. Such a masking of the olefin avoids a possible competing vinylcyclopropane rearrangement. The correspondence of the cyclopentene of *202* with the vinylcyclopropane in *203* now becomes obvious. The presence of the dimethylcarbinol side chain now also offers the opportunity for its introduction by addition of a cyclopropyl anion to acetone. The feasibility of creating such an anion by fluoride initiated desilylation

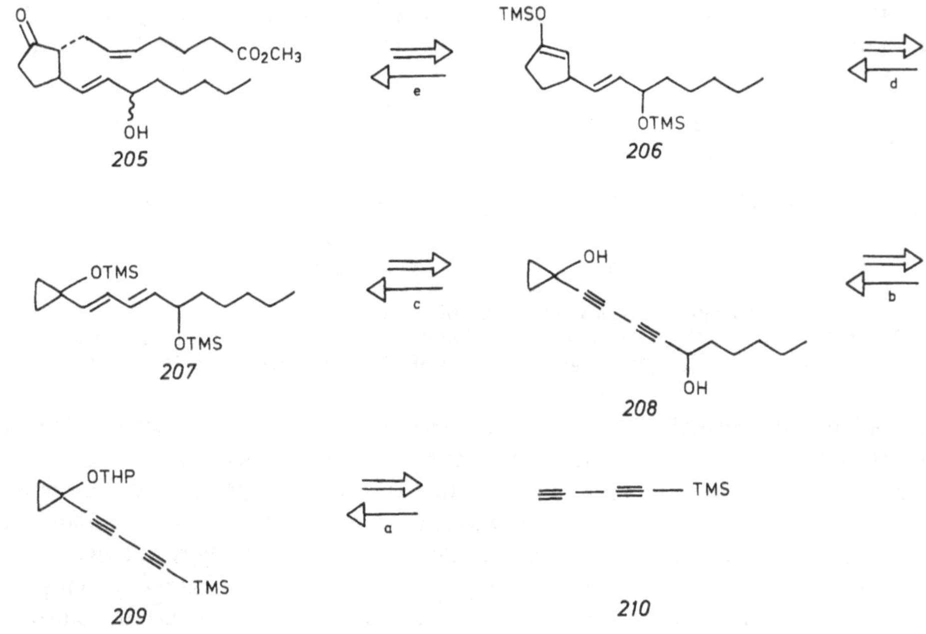

Scheme 12. Retrosynthetic Analysis and Synthesis of 11-Deoxyprostaglandin E
[a] C_4H_9Li, CH_3MgI then *58* followed by DHP, PPTS, 67%; [b] KF, H_2O, DMF then nC_4H_9Li, THF, n-$C_5H_{11}CHO$ followed by C_2H_5OH, PPTS, 61%; [c] LAH, THF then TMS-Cl, $(C_2H_5)_3N$, DMSO, 72%; [d] FVT, 100%; [e] $LiNH_2$, liq. NH_3, $BrCH_2CH{=}CHCH_2CH_2{-}CH_2{-}CO_2CH_3$, 45%

of a trimethylsilylcyclopropane simplifies the synthesis to *204*. While a route to *204* akin to that used in Scheme 10 can be envisioned, a daring more direct alternative would be the cross reductive coupling between 2,6-dimethylcyclohex-2-enone and 1-trimethylsilylcyclopropane-carboxaldehyde. Surprisingly, the low valent titanium mediated reductive olefination using excess aldehyde gave the vinyl cyclopropane in 50–60% yield. Prior to rearrangement, the free hydroxyl group had to be masked as the tertiary ether. Reasonable diastereoselectivity accompanied the vinylcyclopropane rearrangement in which *202* predominated to the extent of 5:1. A scant five steps constituted the entire synthesis.

The hemiketal of cyclopropanone provided an efficient convergent strategy to 11-deoxyprostaglandin E (*205*) as summarized in Scheme 12 [103]. The synthesis of a 2,3-disubstituted cyclopentanone can arise by regioselective alkylation of a 2-substituted cyclopentanone via a regiocontrolled formation of an enolate equivalent as in *206*. Formation of a cyclopentene via a vinylcyclopropane rearrangement factors the problem to a synthesis of *207*. While formation of the latter from an oxaspiropentane is possible, an alternative envisions a nucleophilic addition to cyclopropanone. The sensitivity of butadienyl Grignard reagents suggests a more stable acetylide anion as an alternative. Indeed, if both trans double bonds are transformed into acetylenes as in *208*, introduction of the C(15)–C(20) unit by addition of an acetylide derived from *209* to hexanal also becomes feasible. The penultimate starting materials then become diacetylene as in *210* and cyclopropanone hemiketal. Performing the sequence in the synthetic direction provides the PG derivative in 13% overall yield in only 9 steps.

The heteroatom version of the vinylcyclopropane rearrangement serves to facilitate alkaloid construction. Scheme 13 outlines a strategy for the pyrrolizidine alkaloid isoretronecanol *211* [90]. Use of a carboxaldehyde (i.e. *213*) as a synthon for the primary alcohol provides an ability to adjust stereochemistry. It also sets up formation of the pyrrolidine ring bearing the aldehyde by an aldol-type condensation of an enol of the aldehyde onto an imine derived from *214*. Because of the lability of such systems, introduction of X=PhS imparts stability. The resultant "azacyclopentene" translates to an imine *215* using the iminocyclopropane rearrangement methodology. Simple condensation of the primary amine *216* with aldehyde *37a* then initiates this

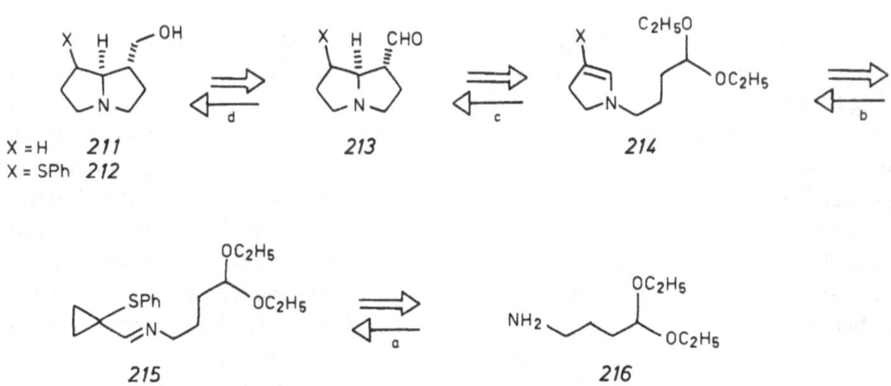

Scheme 13. Retrosynthetic Analysis and Synthesis of Isoretronecanol
[a] *37a*; [b] NH_4Cl, xylene; [c] CH_3OH, HCl; [d] LAH then Raney Ni (No yields)

route. The sulfur substituent also facilitates the rearrangement. The execution of this strategy requires only five steps. The same notion translates into completed synthesis of alkaloids including δ-coniceine, ipalbidine, and septicine [140].

15 Cyclohexyl Synthesis

The extrapolation of the vinylcyclopropane-cyclopentene rearrangement to a vinylcyclobutane-cyclohexene synthesis begins to create new insights into the synthesis of six membered ring natural products. The eudesmane sesquiterpene (—)-β-selinene, *217* illustrates such a strategy as summarized in Scheme 14 [80]. A suitable cyclohexene

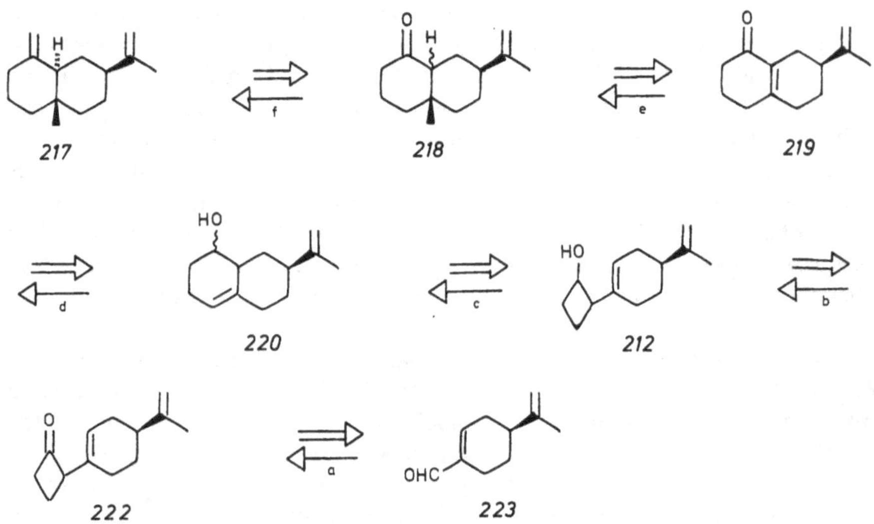

Scheme 14. Retrosynthetic Analysis and Synthesis of (—)-β-Selinene
[a] *34* then HBF_4, H_2O, 67%; [b] LAH, THF; [c] KH, THF; [d] CrO_3, H_2O, H_2SO_4, CH_3COCH_3, 67% overall for steps [b], [c], and [d]; [e] $CH_3Cu \cdot BF_3$, ether, 47%; [f] Ph_3PCH_2, DMSO, 92%

must be created for this strategy. Recognizing the relationship of an exocyclic methylene group and a carbonyl group and that methylenation of 1-decalones adjusts the ring fusion to trans concommitant with olefination generates *218* as the immediate precursor. The presence of a methyl group β to a ketone permits its introduction via conjugate addition to the enone *219* which possesses only a single stereogenic center. Axial attack assures the stereochemistry. For the oxygen assisted vinylcyclobutane rearrangement, a homoallylic alcohol is necessary. Conceptually, simply deconjugating the double bond as in *220* creates the pattern to reorganize the framework to the vinylcyclobutanol *221*. The remaining steps to cyclobutanone *222* and aldehyde *223* then become obvious. The identification of perillaldehyde *223* as the starting material solves the problem of the absolute stereochemistry as well. All the diastereomeric intermediates *222* → *220* which exist as mixtures can be taken on without purification to *219* which then is a homogeneous substance. The six step synthesis proceeds in 19% overall yield. Adding this new dimension to six membered ring synthesis provides additional flexibility in synthetic design to these immense and important classes of compounds.

16 Trisubstituted Olefin Synthesis

The opening of cyclopropylcarbinols to homoallylic bromides constituted the first use of cyclopropyl compounds for the stereocontrolled synthesis of natural products. The cyclopropyl conjunctive reagents enhance the richness of this notion. The stereocontrolled opening of vinylcyclopropanes by a homopentadienyl proton shift provides an approach to trisubstituted olefins and thereby a new strategy. The fungal prohormone methyl trisporate B (*224*) as summarized in Scheme 15 illustrates this conceptual development [97].

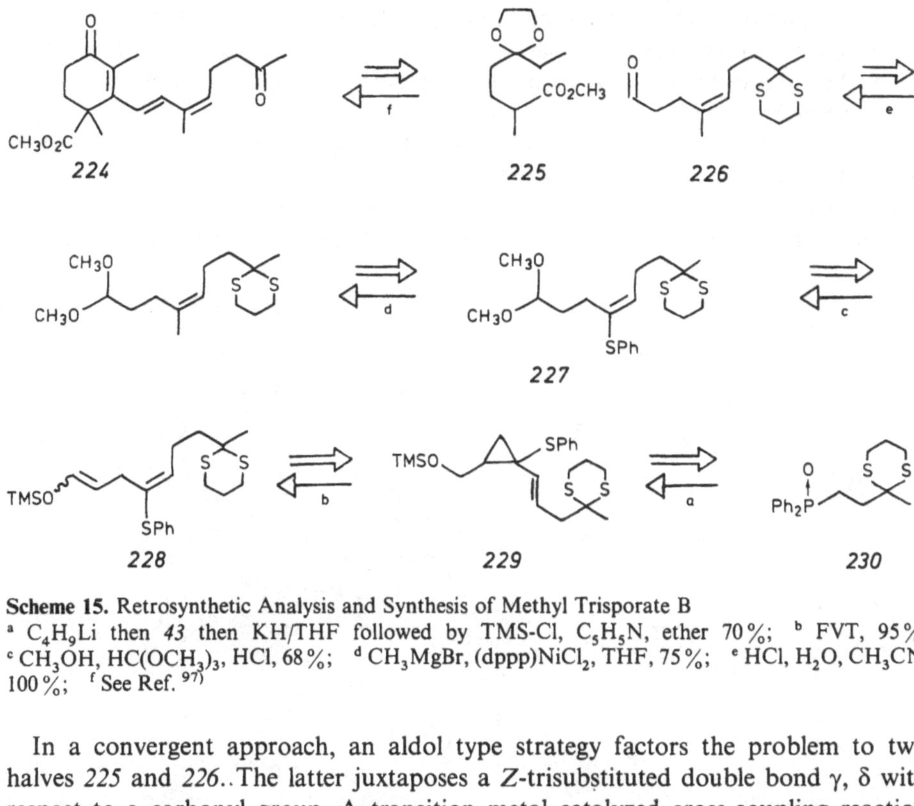

Scheme 15. Retrosynthetic Analysis and Synthesis of Methyl Trisporate B
[a] C_4H_9Li then *43* then KH/THF followed by TMS-Cl, C_5H_5N, ether 70%; [b] FVT, 95%; [c] CH_3OH, $HC(OCH_3)_3$, HCl, 68%; [d] CH_3MgBr, (dppp)$NiCl_2$, THF, 75%; [e] HCl, H_2O, CH_3CN, 100%; [f] See Ref. [97]

In a convergent approach, an aldol type strategy factors the problem to two halves *225* and *226*. The latter juxtaposes a *Z*-trisubstituted double bond γ, δ with respect to a carbonyl group. A transition metal catalyzed cross-coupling reaction permits the replacement of the methyl group by a heteroatom substituent such as phenylthio as in *227*. Since the homopentadienyl rearrangement creates a 1,5-diene, conversion of *227* into such a structure envisions an enol derivative of the aldehyde as in *228*. The vinylcyclopropane then becomes *229* which can derive by olefination of the cyclopropyl conjunctive reagent *43* with the metalated phosphine oxide derived from *230*. The high geometrical control in the homopentadienyl rearrangement of *229* and in the nickel catalyzed cross coupling of *227* enhances the value of this trisubstituted olefin synthesis. By using a cross-coupling process to introduce one of the three olefin substituents, greater versatility exists. This strategy complements the Julia method by producing the thermodynamically less stable *Z* isomer; whereas, the Julia method produces the thermodynamically more stable *E* isomer.

17 Conclusions

The unique bonding features associated with small ring compounds seems ideal to meet the demands for selective transformations. The introduction of units of high potential energy provides a powerful force for release of strain. The availability for generating a whole family of compounds by introduction of various substituents can not only affect the substitution pattern of the resultant products but can also serve as steering groups to direct the reaction along different paths. While the complexity of the molecular change accompanying the unraveling of the cyclopropyl ring makes the connection between target and precursor more obtuse, the efficiency that can result makes the efforts well worthwhile.

Some of the existing reagents offer unusual stereochemical control. In several cases, the reactivity of these reagents allowed goals to be reached that were unapproachable. The diversity offered by such reagents as exemplified by the varied structural units available from the nucleophilic conjunctive cyclopropyl reagents as

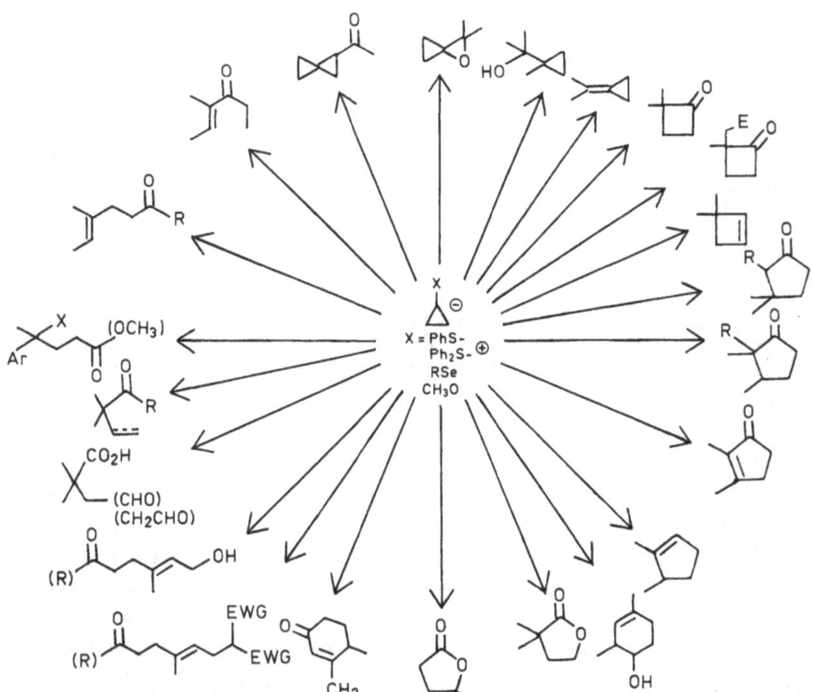

Fig. 1. Structural Units Available with Nucleophilic Cyclopropyl Reagents

summarized in Fig. 1 stimulates development of other such reagents. Further understanding of the reactivity of small ring compounds will undoubtedly lead to additional synthetic applications of existing reagents. Combined with the creation of untold numbers of new small ring conjunctive reagents, the promise for future developments in immense.

18 References

1. Pell, A. S., Pilcher, G.: Trans. Faraday Soc. *61*, 71 (1965)
2. Doering, W. von E., LaFlamme, P. M.: Tetrahedron *2*, 75 (1958); Moore, W. R., Ward, H. R.: J. Org. Chem. *27*, 4179 (1962); Skattebol, L.: Acta Chem. Scand. *17*, 1683 (1963). For a review see Hopf, H.: in The Chemistry of Ketenes, Allenes, and Related Compounds, Part 1, (ed. Patai, S.) John Wiley and Sons, Chichester, 1980, pp. 779–902
3. Giese, B., Zwick, W.: Chem. Ber. *115*, 2526 (1982). Also see Levina, R. Ya., Bostin, B. N., Ustynyuk, T. K.: J. Gen. Chem. USSR *30*, 383 (1960); Hensen, F. R., Patterson, D. B., Dinizo, S. E.: Tetrahedron Lett. *1974*, 1315
4. Wenkert, E., Berges, D. A., Golob, N. R.: J. Am. Chem. Soc. *100*, 1263 (1978)
5. Johnson, W. S., Ti, T., Faulkner, D. J., Campbell, S. F.: ibid. *90*, 6225 (1968); Brady, S. F., Ilton, M. A., Johnson, W. S.: ibid. *90*, 2882 (1968); Julia, M., Julia, S., Guegan, R.: Bull. Soc. Chim. France 1072 (1960)
6. a) Knapp, S., Trope, A. F., Theodore, M. S., Hirata, N., Barchi, J. J.: J. Org. Chem. *49*, 608 (1984); b) Ogura, K., Yamashita, M., Suzuki, M., Tsuchihashi, G.: Chem. Lett. *1982*, 93. Also see Abe, K., Okumura, H., Tsugoshi, T., Nakamura, N.: Synthesis 603 (1984)
7. For some other recent examples of ring expansions see Depres, J. P., Greene, A. E.: J. Org. Chem. *45*, 2036 (1980); Mehta, G., Rao, K. S.: Tetrahedron Lett. *25*, 1839 (1984); Leriverend, M. L., Leriverend, P.: Chem. Ber. *109*, 3492 (1976)
8. Bogdanowicz, M. J., Ambelang, T., Trost, B. M.: Tetrahedron Lett. 923 (1973); Tsuda, Y., Tanno, T., Ukai, A., Isobe, K.: ibid. 2009 (1971)
9. Ghosez, L., Montaigne, R., Roussel, A., van Lierde, H., Mollet, P.: Tetrahedron *27*, 615 (1971); Terlinden, R., Boland, W., Jaenicke, L.: Helv. Chim. Acta *66*, 466 (1983)
10. Cossement, E., Biname, R., Ghosez, L.: Tetrahedron Lett. 997 (1974); Ishida, M., Minami, T., Agawa, T.: J. Org. Chem. *44*, 2067 (1979); Michel, P., O'Donnell, M., Biname, R., Hesbain-Frisque, A. M., Ghosez, L., DeClercq, J. P., Germain, G., Arte, E., van Meerssche, M.: Tetrahedron Lett. *21*, 2577 (1980)
11. Brule, D., Chalchat, J.-C., Garry, R.-P., Lacroix, G., Michet, A., Vessiere, R.: Bull. Soc. Chim. France II–57 (1981)
12. Trost, B. M., Scudder, P. H.: J. Org. Chem. *46*, 506 (1981)
13. Danheiser, R.: presented at Internat. Symp. on Heteroatoms for Organic Synthesis, Montreal, August 1983
14. Glass, D. S., Boikess, R. S., Winstein, S.: Tetrahedron Lett. 999 (1966)
15. Hudlicky, T., Koszyk, F. J., Kutchan, T. M., Sheth, J. P.: J. Org. Chem. *45*, 5020 (1980)
16. Ohloff, G., Pickenhagen, W.: Helv. Chim. Acta *52*, 880 (1969); Marino, J. P., Kaneko, T.: J. Org. Chem. *39*, 3175 (1974); Piers, E., Reudiger, E. H.: Chem. Commun. 166 (1979); Wender, P. A., Eissenstat, M. A., Filosa, M. P.: J. Am. Chem. Soc. *101*, 2196 (1979); Wender, P. A., Hillemann, C. L., Szymonifka, M. J.: Tetrahedron Lett. *21*, 2205 (1980); Piers, E., Jung, G. L., Moss, N.: Tetrahedron Lett. *25*, 3959 (1984)
17. Klumpp, G. W., Kool, M., Schakel, M., Schmitz, R. F., Bouthan, C.: J. Am. Chem. Soc. *101*, 7065 (1979)
18. Thomas, E. W.: Tetrahedron Lett. *24*, 2347 (1983)
19. Piers, E., Lau, C. K., Nagakura, I.: Tetrahedron Lett. 3233 (1976); Piers, E., Banville, J., Lau, C. K., Nagakura, I.: Can. J. Chem. *60*, 2965 (1982)
20. Piers, E., Nagakura, I.: Tetrahedron Lett. 3237 (1976)
21. Marino, J. P., Browne, L. J.: ibid. 3241, 3245 (1976)
22. Sugihara, Y., Wakabayashi, S., Murahata, I.: J. Am. Chem. Soc. *105*, 6718 (1983)

23. Corey, E. J., Ulrich, P.: Tetrahedron Lett. 3685 (1975)
24. Morgans, D. J., Jr., Feigelson, G. B.: J. Am. Chem. Soc. *105*, 5477 (1983)
25. For reviews see: a) Trost, B. M.: Gazz. Chim. Ital. *114*, 139 (1984); b) Pure Appl. Chem. *43*, 563 (1975); c) Organic Sulfur Chemistry (Stirling, C. J. M. ed.), Butterworths, London, 1975, pp. 237–264; d) Accounts Chem. Res. *7*, 85 (1974)
26. Bogdanowicz, M. J., Trost, B. M.: Org. Syn. *54*, 27 (1974): For an alternative non-silver based preparation of the acyclic precursor see: Badet, B., Julia, M.: Tetrahedron Lett. 1101 (1979)
27. Trost, B. M., Bogdanowicz, M. J.: J. Am. Chem. Soc. *95*, 5298 (1973)
28. Johnson, C. R., Janiga, E. R.: ibid. *95*, 7692 (1973)
29. a) Bogdanowicz, M., Trost, B. M.: Tetrahedron Lett. 887 (1972); b) Trost, B. M., Bogdanowicz, M. J.: J. Am. Chem. Soc. *95*, 5311 (1973)
30. Trost, B. M., Bogdanowicz, M. J.: J. Am. Chem. Soc. *95*, 5307 (1973)
31. Also see: Seebach, D., Dammann, R., Lindner, H. J., Kitschke, B.: Helv. Chim. Acta *62*, 1143 (1979)
32. Braun, M., Dammann, R., Seebach, D.: Chem. Ber. *108*, 2368 (1975)
33. Dammann, R., Braun, M., Seebach, D.: Helv. Chim. Acta *59*, 2821 (1976)
34. Collins, C. J., Hanack, M., Stutz, H., Auchter, G., Schoberth, W.: J. Org. Chem. *48*, 5260 (1983)
35. Trost, B. M., Bogdanowicz, M. J.: J. Am. Chem. Soc. *95*, 5321 (1973)
36. Trost, B. M., Bogdanowicz, M. J., Frazee, W. J., Salzmann, T. N.: ibid. *100*, 5512 (1978)
37. Trost, B. M., Scudder, P. H.: ibid. *99*, 7601 (1977)
38. Feiner, N. F., Abrams, G. D., Yates, P.: Can. J. Chem. *54*, 3955 (1976)
39. a) Trost, B. M., Rigby, J. H.: J. Org. Chem. *41*, 3217 (1976); b) Rigby, J. H.: Ph.D. Thesis, Univ. Wisconsin 1977
40. Trost, B. M., Bogdanowicz, M. J., Kern, J.: J. Am. Chem. Soc. *97*, 2218 (1975)
41. Miller, D. D., Bossart, J. F., Chalekis, K.: J. Org. Chem. *44*, 4449 (1979)
42. Mao, M. K.-T.: Ph.D. Thesis, Univ. Wisconsin 1980
43. Trost, B. M., Balkovec, J. M., Mao, M. K.-T.: J. Am. Chem. Soc. *105*, 6755 (1983)
44. Lightner, D. A., Jackman, D. E.: Chem. Commun. 334 (1974)
45. Frazee, W. J.: Ph.D. Thesis, Univ. Wisconsin 1977
46. Trost, B. M., Frazee, W. J.: J. Am. Chem. Soc. *99*, 6124 (1977)
47. Trost, B. M., Hiroi, K., Holy, N.: ibid. *97*, 5873 (1975)
48. Stanton, J. L.: Ph.D. Thesis, Univ. Wisconsin 1975
49. Trost, B. M., Kurozumi, S.: Tetrahedron Lett. 1929 (1974); unpublished work
50. Bernaert, E., Danneels, D., Anteunis, M., Verhegge, G.: Tetrahedron *29*, 4127 (1973)
51. Lightner, D. C., Chang, T. C.: J. Am. Chem. Soc. *96*, 3015 (1974)
52. Baldwin, J. E., Beckwith, P. L. M.: Chem. Commun. 279 (1983)
53. Green, M. J., Shue, H.-J.: U.S. Patent 4079054; Chem. Abstr. *89*, 110124p (1978)
54. Green, M. J., Shue, H.-J., Shapiro, E. L., Gentles, M. A.: U.S. Patent 4076 708; Chem. Abstr. *89*, 110119r (1978)
55. Trost, B. M., Latimer, L. H.: J. Org. Chem. *43*, 1031 (1978)
56. Trost, B. M., Bogdanowicz, M. J.: J. Am. Chem. Soc. *93*, 3773 (1971)
57. Salaun, J. R., Conia, J. M.: Chem. Commun. 1579 (1971); Salaun, J., Garnier, B., Conia, J. M.: Tetrahedron *30*, 1413 (1974). Also see: Aue, D. H., Meshishnek, M. J., Shellhamer, D. F.: Tetrahedron Lett. 4799 (1973)
58. Green, M. J., Shue, H.-J., McPhail, A. T., Miller, R. W.: Tetrahedron Lett. 2677 (1976)
59. Utimoto, K., Tamura, M., Sisido, K.: Tetrahedron *29*, 1169 (1973); Sisido, K.: Tetrahedron Lett. *28*, 3267 (1968); Schweizer, E. E.: J. Org. Chem. *33*, 337 (1968); Bestmann, H. J., Hartung, H., Pils, I.: Angew. Chem. Internat. Ed. Engl. *4*, 957 (1965); Noyori, R., Takaya, H., Nakanisi, Y., Nozaki, H.: Gan. J. Chem. *47*, 1242 (1969); Chesick, J. P.: J. Am. Chem. Soc. *85*, 2720 (1963)
60. Fitjer, L., Giersig, M., Clegg, W., Schormann, N., Sheldrick, G. M.: Tetrahedron Lett. *24*, 5351 (1983)
61. See ref. 29 and references cited therein
62. Trost, B. M., Bogdanowicz, M. J.: J. Am. Chem. Soc. *95*, 289 (1973)
63. Scudder, P. H.: Ph.D. Thesis, Univ. Wisconsin 1977

64. Trost, B. M., Brandi, A.: J. Am. Chem. Soc. *106*, 5041 (1984)
65. Trost, B. M., Nishimura, Y., Yamamoto, K.: ibid. *101*, 1328 (1979)
66. Trost, B. M., Keeley, D., Bogdanowicz, M. J.: ibid. *95*, 3068 (1973)
67. Tanaka, K., Unema, H., Matsui, S., Kaji, A.: Bull. Chem. Soc. Japan *55*, 2965 (1982)
68. Cohen, T., Daniewski, W. M., Weisenfeld, R. B.: Tetrahedron Lett. 4665 (1978); Cohen, T., Matz, J. R.: Syn. Commun. *10*, 311 (1980)
69. Cohen, T., Daniewski, W. M.: Tetrahedron Lett. 2991 (1978)
70. Halazy, S., Lucchetti, J., Krief, A.: ibid. 3971 (1978)
71. Cohen, T., Matz, J. R.: J. Am. Chem. Soc. *102*, 6900 (1980)
72. Trost, B. M., Vladuchick, W. C.: Synthesis 821 (1978)
73. Trost, B. M., Keeley, D. E., Arndt, H. C., Rigby, J. H., Bogdanowicz, M. J.: J. Am. Chem. Soc. *99*, 3080 (1977)
74. Trost, B. M., Keeley, D. E., Arndt, H. C., Bogdanowicz, M. J.: ibid. *99*, 3088 (1977)
75. Trost, B. M., Rigby, J. H.: J. Org. Chem. *43*, 2938 (1978)
76. Miller, R. D., McKean, D. R.: Tetrahedron Lett. 583 (1979)
77. Halazy, S., Krief, A.: Chem. Commun. 1136 (1979); Denis, J. N., Krief, A.: ibid. 229 (1983)
78. Gadwood, R. C., Lett, R. M.: J. Org. Chem. *47*, 2268 (1982)
79. Greenlee, M. L.: J. Am. Chem. Soc. *103*, 2425 (1981)
80. Cohen, T., Bhupathy, M., Matz, J. R.: ibid. *105*, 520 (1983)
81. Miller, R. D.: Chem. Commun. 277 (1976)
82. Miller, R. D., McKean, D. R., Kaufmann, D.: Tetrahedron Lett. 587 (1979)
83. Cohen, T., Matz, J. R.: ibid. *22*, 2455 (1981)
84. Lyle, T. A., Frei, B.: Helv. Chim. Acta *64*, 2598 (1981)
85. In a private communication, Prof. Krief has informed us that subsequent experiments do not support his early claim (ref. 70) of higher reactivity for the selenium analogue
86. Trost, B. M., Keeley, D. E.: J. Am. Chem. Soc. *96*, 1252 (1974)
87. Halazy, S., Krief, A.: Tetrahedron Lett. *22*, 1829 (1981)
88. Trost, B. M., Jungheim, L. N.: J. Am. Chem. Soc. *102*, 7910 (1980)
89. Halazy, S., Krief, A.: Tetrahedron Lett. *22*, 1833 (1981)
90. Stevens, R. V., Luh, Y., Sheu, J.-T.: ibid. 3799 (1976)
91. Trost, B. M., Ornstein, P. L.: J. Org. Chem. *47*, 748 (1982)
92. Danheiser, R. L., Martinez-Davila, C., Sard, H.: Tetrahedron *37*, 3943 (1981)
93. Ollivier, J., Salaun, J.: Tetrahedron Lett. *25*, 1269 (1984). A benzylic Wittig reagent also is reported to give an *E,Z*-mixture with *39*: Salaun, J., Almirantis, Y.: Tetrahedron *39*, 2421 (1983)
94. Ornstein, P. L.: Ph.D. Thesis, Univ. Wisconsin 1982
95. Trost, B. M., Ornstein, P. L.: J. Org. Chem. *48*, 1131 (1983)
96. Barnier, J. P., Salaun, J.: Tetrahedron Lett. *25*, 1273 (1984)
97. Trost, B. M., Ornstein, P. L.: ibid. *24*, 2833 (1983)
98. a) Salaun, J.: Chem. Rev. *83*, 619 (1983); b) Wasserman, H. H., Clark, G. M., Turley, P. C.: Top. Curr. Chem. *47*, 73 (1974)
99. a) Wasserman, H. H., Clagett, D. C.: J. Am. Chem. Soc. *88*, 5368 (1966); b) Salaun, J.: J. Org. Chem. *41*, 1237 (1976)
100. Salaun, J.: ibid. *42*, 28 (1977)
101. a) Wasserman, H. H., Hearn, M. J., Cochoy, R. E.: ibid. *45*, 2874 (1980); b) Wasserman, H. H., Cochoy, R. E., Baird, M. S.: J. Am. Chem. Soc. *91*, 2375 (1969)
102. Salaun, J., Bennani, F., Compain, J. C., Fadel, A., Ollivier, J.: J. Org. Chem. *45*, 4129 (1980)
103. Salaun, J., Ollivier, J.: Nouveau J. Chim. *5*, 587 (1981)
104. Salaun, J., Garnier, B., Conia, J. M.: Tetrahedron *30*, 1413 (1974)
105. Trost, B. M., Mao, M. K.: J. Am. Chem. Soc. *105*, 6753 (1983)
106. Trost, B. M., Brandi, A.: ibid. *106*, 5041 (1984)
107. Eaton, P. E., Carlson, G. R., Lee, J. T.: J. Org. Chem. *38*, 4071 (1973)
108. Greene, A. E., Depres, J. P., Nagano, H., Crabbe, P., Crabbe, P.: Tetrahedron Lett. 2365 (1977)
109. Depres, J. P., Greene, A. E.: J. Am. Chem. Soc. *101*, 4003 (1979)

110. Gadwood, R. C.: J. Org. Chem. *48*, 2098 (1983)
111. Cf.: Halazy, S., Zutterman, F., Krief, A.: Tetrahedron Lett. *23*, 4385 (1982); Halazy, S., Krief, A.: Chem. Commun. 1200 (1982)
112. Matz, J. R., Cohen, T.: Tetrahedron Lett. *22*, 2459 (1981)
113. Trost, B. M., Keeley, D. E.: J. Am. Chem. Soc. *98*, 248 (1976)
114. Keeley, D. E.: Ph.D. Thesis, Univ. Wisconsin 1976
115. Paquette, L. A., Horn, K. A., Wells, G. J.: Tetrahedron Lett. *23*, 159 (1982); Paquette, L. A., Horn, K. A., Wells, G. J., Yan, T.-H.: ibid. *23*, 263 (1982)
116. Paquette, L. A., Horn, K. A., Wells, G. J., Yan, T.-H.: Tetrahedron *39*, 913 (1983)
117. Wasserman, H. H., Dion, R. P.: Tetrahedron Lett. *24*, 3409 (1983); *23* 1413 (1982)
118. Bertrand, M. E., Gil, G., Junino, A., Maurin, R.: J. Chem. Res. 1551 (1980)
119. Bhupathy, M., Cohen, T.: J. Am. Chem. Soc. *105*, 6978 (1983)
120. Kahn, M.: Tetrahedron Lett. *21*, 4547 (1980)
121. Levine, S. G., McDaniel, R. L., Jr.: J. Org. Chem. *46*, 2199 (1981)
122. Miller, R. D., McKean, D. R.: Tetrahedron Lett. *20*, 1003, 587, 583 (1979)
123. Miller, R. D., McKean, D. R.: ibid. *21*, 2639 (1980)
124. Trost, B. M., Bogdanowicz, M. J.: J. Am. Chem. Soc. *94*, 4777 (1972)
125. Trost, B. M., Bogdanowicz, M. J.: ibid. *95*, 2038 (1973)
126. Also see: Conia, J. M., Ripoll, J. L.: Bull. Chim. Soc. France 763 (1963); Ghosez, L., Montaigne, R., Roussel, A., van Lierde, H., Molliet, P.: Tetrahedron *27*, 615 (1971)
127. Conia, J. M., Salaun, J. R.: Acc. Chem. Res. *5*, 33 (1972); Conia, J. M., Robson, M. J.: Angew. Chem. Internat. Ed. Engl. *14*, 473 (1975)
128. Trost, B. M., Preckel, M.: J. Am. Chem. Soc. *95*, 7862 (1973)
129. Trost, B. M., Preckel, M., Leichter, L. M.: ibid. *97*, 2224 (1975)
130. Also see: Cossement, E., Biname, E., Ghosez, L.: Tetrahedron Lett. 997 (1974)
131. Marshall, J. A., Seitz, D. E.: J. Org. Chem. *39*, 1814 (1974)
132. Bordwell, F. G.: Pure Appl. Chem. *49*, 963 (1977)
133. Morizawa, Y., Kanakura, A., Yamamoto, H., Hiyama, T., Nozaki, H.: Bull. Chem. Soc. Japan *57*, 1935 (1984)
134. Trost, B. M., Keeley, D. E.: J. Org. Chem. *40*, 2013 (1975)
135. a) Balkovec, J. M.: unpublished work; b) Trost, B. M., Balkovec, J. M.: Tetrahedron Lett. *26*, 0000 (1985); c) Allamandin: Kupchan, S. M., Dessertine, A. L., Blaylock, B. T., Bryan, R. F.: J. Org. Chem. *39*, 2477 (1974); d) Plumericin: Albers-Schonberg, G., Schmid, G.: Helv. Chim. Acta *44*, 1447 (1961); e) Abe, F., Mori, T., Yamauchi, T.: Chem. Pharm. Bull. *32*, 2947 (1984)
136. Paquette, L.: Proc. Natl. Acad. Sci. USA *79*, 4495 (1982)
137. Paquette, L. A., Wyvratt, M. J., Schallner, O., Muthard, J. L., Begley, W. J., Blankenship, R. M., Balogh, D.: J. Org. Chem. *44*, 3616 (1979)
138. Dalziel, W., Hesp, B., Stevenson, K. M., Jarvis, J. A.: J. Chem. Soc. Perkin Trans 1, 2841 (1973)
139. Yan, T. H., Paquette, L. A.: Tetrahedron Lett. *23*, 3227 (1982)
140. Stevens, R. V., Luh, Y.: ibid. 979 (1977)

The Application of Cyclobutane Derivatives in Organic Synthesis

Henry N. C. Wong[1], Kin-Lik Lau[2] and Kam-Fai Tam[2]

1 Department of Chemistry, The Chinese University of Hong Kong, Shatin, New Territories, Hong Kong.
2 Department of Applied Science, Hong Kong Polytechnic, Hung Hom, Kowloon, Hong Kong

A great variety of methods are available for the synthesis of cyclobutane derivatives. These methods generally allow regioselective as well as stereoselective synthesis of extensively substituted four-membered ring carbocycles. Although possessing intense ring strain, cyclobutanes are normally stable at room temperature, hence can be handled easily in laboratories. Nevertheless, four-membered ring compounds can undergo ring cleavage with extreme ease under several reaction conditions such as acidic condition, basic condition, nucleophilic attack, thermolysis, photolysis, oxidizing as well as reducing conditions. When the cyclobutane ring under investigation is appropiately designed to link to other functional groups, ring opening can be usually followed by skeletal rearrangement. These intriguing reactions would yield compounds with very complex structures. Therefore, the combined effect of all these special properties enables cyclobutane derivatives to become very versatile and useful starting materials for organic synthesis.

1 Introduction

The construction of molecular frameworks and the control of stereochemistry thereon remain as great challenges to synthetic organic chemists. Nowadays, numerous research teams are engaging in devising simple methodology in order to prepare extremely complex natural molecules. Amongst these many newly developed methods, most noteworthy is the application of cyclobutane derivatives. This approach has emerged as a powerful method which enables organic chemists to synthesize compounds regio-as well as stereoselectively. The unquestionable popularity of cyclobutane derivatives has been attained due mainly to two reasons. In the first instance, four-membered ring compounds are easily accessible. A great number of synthetic methods for making cyclobutanes have been developed in the last three decades [1,2,3]. By using one or the other methods, it is not difficult for an organic chemist to prepare a four-membered ring compound with desired structure and then subject this molecule to cleavage reaction condition. Secondly, cleavage of cyclobutane rings are extremely facile due primarily to their inherent strain [1,3,4]. The ease of cleavage varies also with substituents of the ring, nature of the reagents and conditions of the reaction [3]. Thus, a cyclobutane ring can be fissioned and form noncyclic compounds, fissioned and recyclize to other cyclic compounds such as cyclopentanes, cyclohexanes as well as heterocyclic compounds.

This review would give a brief survey of the literature concerning the synthetic application of cyclobutanes published mainly during 1975 to 1983. Due to the enormity of research efforts in this field, we do not intend to provide a thorough account of this subject, but would only try to present some representative research works which conclusively demonstrate the versatility of cyclobutane derivatives in synthetic organic chemistry.

2 Ring Opening by Acid, Base, Electrophiles or Nucleophiles

2.1 Retro-aldol Reaction

Irradiation of a 12% solution of acetylacetone in cyclohexene afforded the diketone (*2*), which should go through the intermediate (*1*) following a retro-aldol reaction [5,6]. This reaction is best known as the de Mayo reaction [7], which enables the

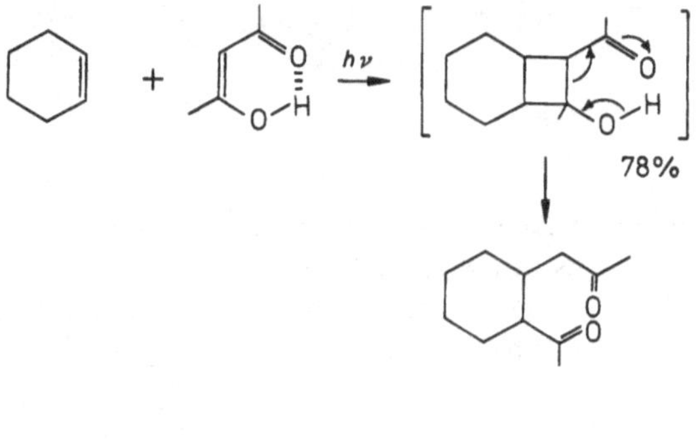

1 (*1*)

2 (*2*)

synthesis of 1,5-diketones by photocycloaddition of enol derivatives of 1,3-diketones to olefins, following by a retro-aldol reaction.

The reaction itself is rather attractive and useful as a synthetic tool, because of its practical simplicity. Indeed, 1,5-dicarbonyl compounds can be readily prepared from olefin and 1,3-dicarbonyl compound in a one-pot reaction. Subsequent aldolization of the resulting 1,5-dicarbonyl compounds would furnish cyclohexenones. The reaction mechanism and stereochemistry of the de Mayo reaction has been comprehensively reviewed [7], and the basic strategy of which has also been developed and extended. The following examples would demonstrate some applications and variations of the de Mayo reaction.

The photo[2 + 2]cycloaddition of alkene (*3*) to the enol (*4*) yielded the diketone (*6*) via the intermediate (*5*) [8]. (Table 1)

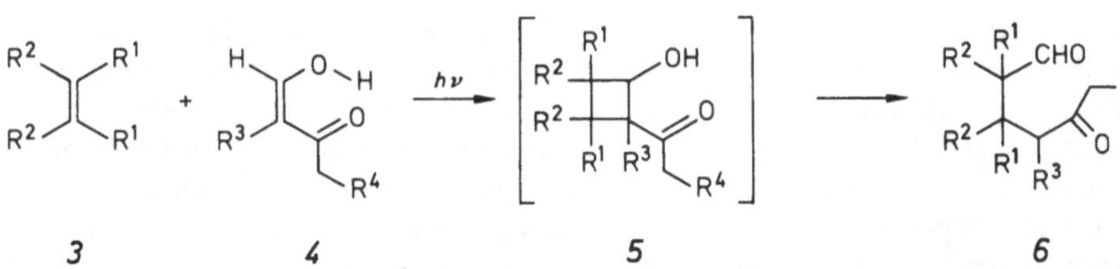

Table 1

R^1	R^2	R^3	R^4	Yield of (*6*)
Me	Me	H	H	100%
Me	Me	H	^{i}Pr	48–90%
H	$-(CH_2)_4-$	H	H	48–90%
H	$-(CH_2)_4-$	H	^{i}Pr	48–90%
H	$-(CH_2)_4-$	Me	Me	48–90%
Me	$-(CH_2)_4-$	H	H	100%
Me	$-(CH_2)_4-$	H	^{i}Pr	75%

Treatment of the cyclobutane (*7*) with acid yielded 72% of the compound (*8*),

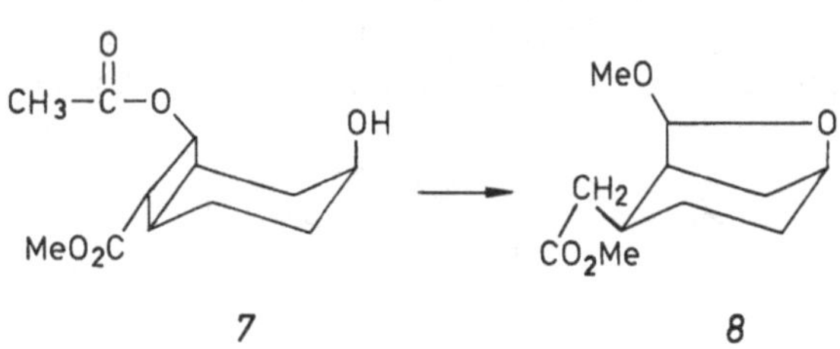

which was suggested to undergo the do Mayo process [9]. Compound (*8*) was a key intermediate in the total synthesis of reserpine [10].

It is noteworthy to point out that the rate of retro-aldol reaction is enhanced by ring strain. Thus, suitably substituted 3-oxygenated carbonyl cyclobutanes would undergo retro-aldol reaction with extreme ease and provide dicarbonyl compounds in relatively high yields. When the molecule has the capability of reacting further, the obtained dicarbonyl would be transformed subsequently to other cyclic systems. This strategy has been applied in the synthesis of very complex molecules and some of the examples will be depicted as follows:

When the acetate (*9*) and (*11*) were treated with acid, retro-aldol, aromatization and dehydration reactions would take place and the final products (*10*) and (*12*) were formed. The structures of (*10*) and (*12*) have been proved to be naphthofurans [11].

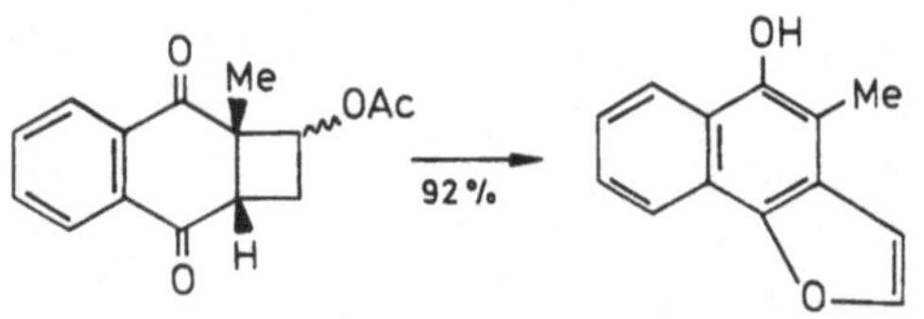

9 *10*

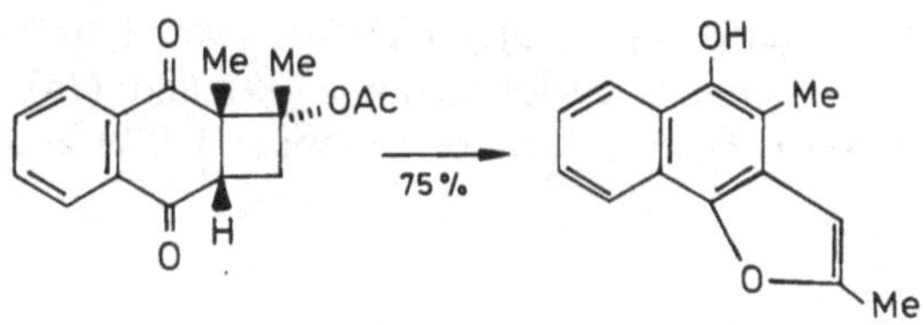

11 *12*

Hydrolysis of the trimethylsiloxy compound (*13*) furnished the diketone (*14*) [12].

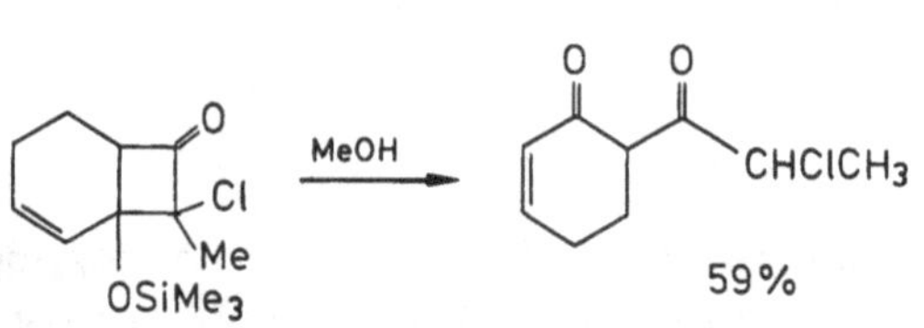

13 *14*

A modification of this work was recently reported by Liu who photolyzed the cycloalkenone (*15*) with ethyl β-trimethylsilyloxy crotonate (*16*) to afford the

Table 2

n	R^1	R^2	R^3	Yield of (*17*)	Yield of (*18*)
0	H	H	H	65%	90%
1	H	H	H	99%	71%
1	H	Me	H	92%	90%
1	Me	H	Me	100%	70%

cyclobutanoid (*17*). Fluoride ion cleaved the 4-membered ring easily to provide the ketoesters (*18*) (Table 2) [13].

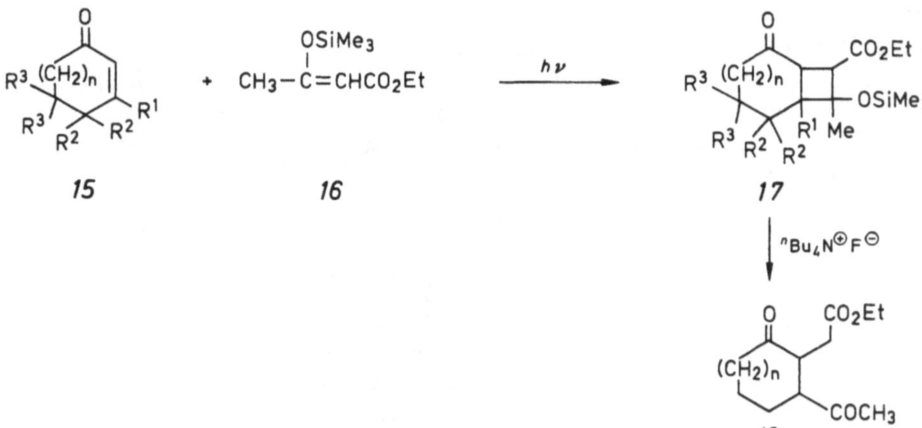

15 *16* *17* *18*

Although the enones (*15*) underwent [2 + 2]cycloaddition with (*16*) with complete regiospecificity, the reaction between cyclohexenone (*19*) and (*16*) yielded the isomeric inseparable (*20*) and (*21*) in 3:1 ratio. Compound (*20*) and (*21*), upon

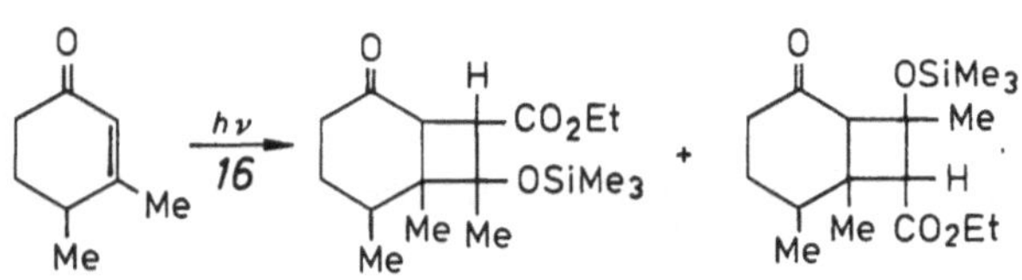

19 *20* *21*

treatment with $^{n}Bu_4N^+F^-$, afforded 50% of the desired (*22*), together with the starting material (*19*) and ethyl acetoacetate (*23*). It was suggested that (*21*) could undergo reaction to yield (*19*) and (*23*) [13]. Similarly, the cycloaddition between 4-cholesten-3-one (*24*) and (*16*) yield the isomeric (*25*) and (*26*) in a ratio of 1:1. Treatment of the inseparable mixture of (*25*) and (*26*) with $^{h}Bu_4N^+F^-$ yielded 40% of (*27*), 40% of (*24*) and also (*23*) [13].

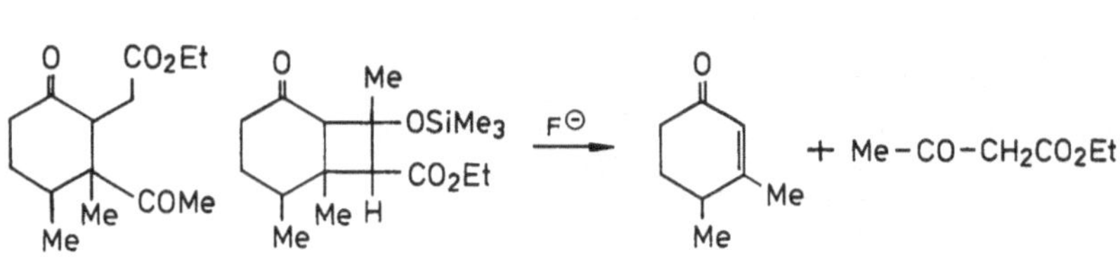

22 *21* *19* *23*

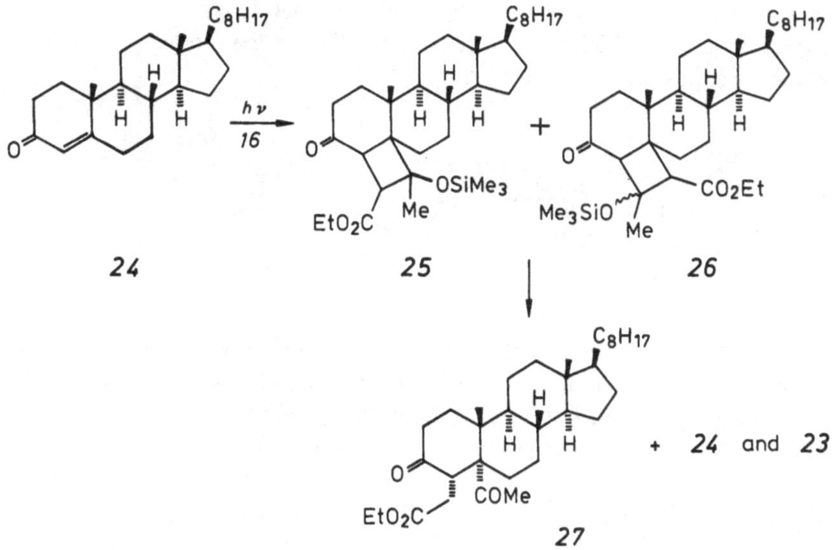

Treatment of the epoxide (*28*) with ethanolic solution of PhSH containing KOH afforded the aldol product (*31*), which was believed to form via the nucleophile promoted retro-aldolization and subsequent aldolization reactions [14].

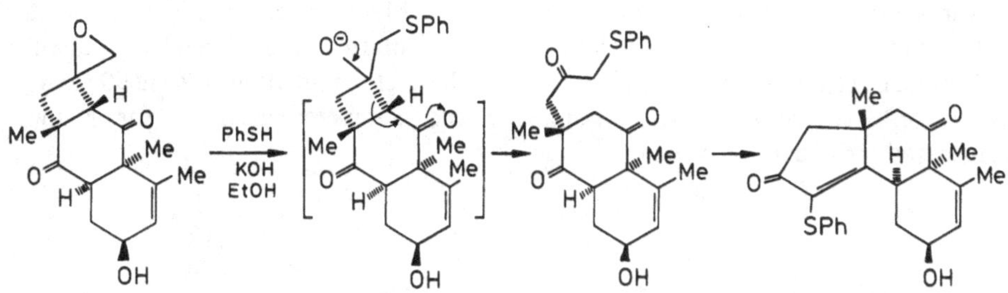

The reduction of (*32*) with $NaBH_4$ also induced the retro-aldol reaction which would give the *syn*-compound (*33*) and finally an epimerization reaction would convert (*33*) to the *anti*-compound (*34*) [1].

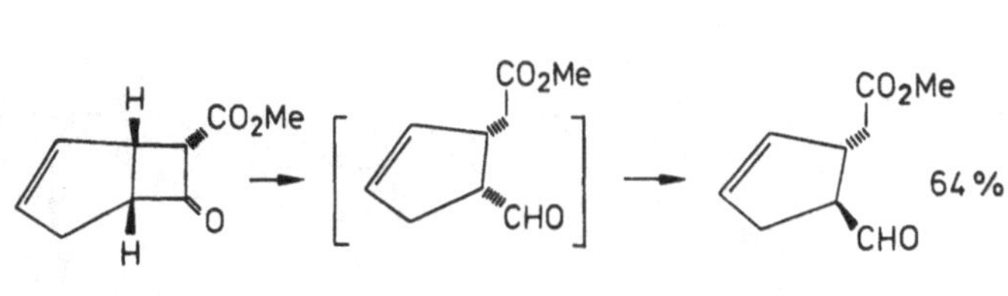

Treatment of the diketone (*35*) with 1N HCl in MeOH gave the crystalline dimethylstipitatonate (*36*) which was the key intermediate in the total synthesis of the mould metabolite stipitotanic acid (*37*) [16].

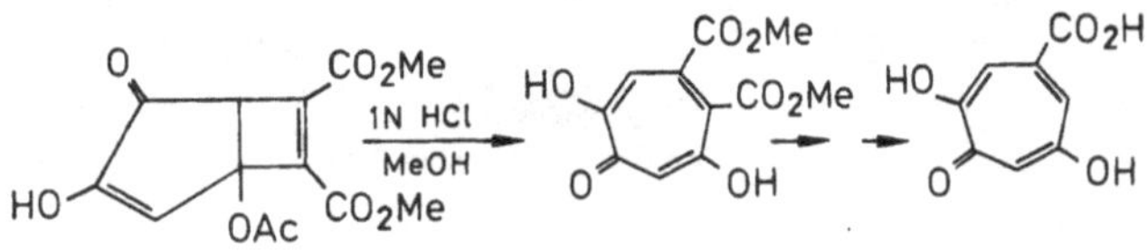

35 36 37

Irradiation of the enol acetate (*38*) gave pure tricyclic acetoxyketone (*39*) in quantitative yield [17,18]. On alkaline hydrolysis, compound (*39*) was converted to the intramolecular de Mayo reaction product (*40*) in 78% yield [17,18]. Epi-precapnelladiene (*41*) was obtained by further reactions of compound (*40*) [19].

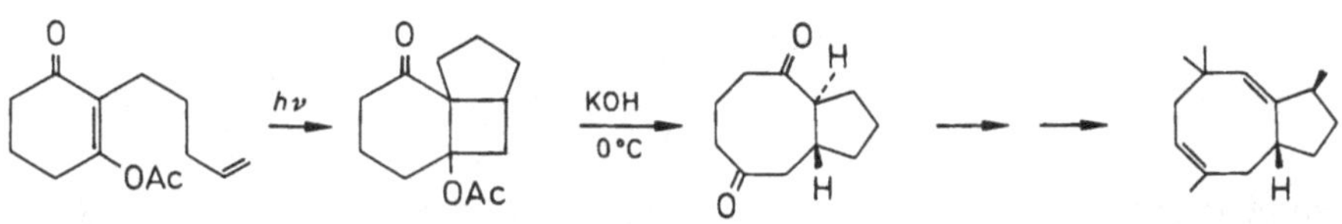

38 39 40 41

Pattenden also made use of this strategy to synthesize $\Delta^{8(9)}$-capnellene [20].

An equilibrium mixture of the cyclohexane-1,3-dione (*42*) and its enol form (*43*) was irradiated in the presence of cyclopentene in MeOH to afford the intermediate (*44*), which was readily transformed to the tricyclic intermediate (*45*) and subsequently followed an retroaldolization sequence to give the cyclooctanedione (*46*) in 90% yield. When refluxed with titanium trichloride and K metal in THF for 5 min., compound (*46*) gave the diol (*47*) [21].

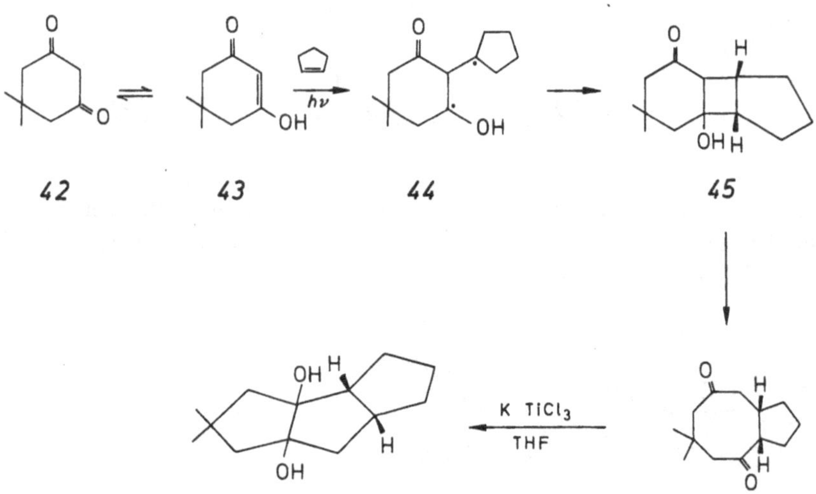

42 43 44 45

47 46

Photoaddition of the enol acetate (*48*) in hexane resulted in the intramolecular [2 + 2]cycloaddition with the production of a 2:3 mixture of the isomeric tricyclic ketones (*49*) and (*50*). Upon hydrolytic cleavage, the mixture was converted to the compounds (*51*) and (*52*) [18, 22].

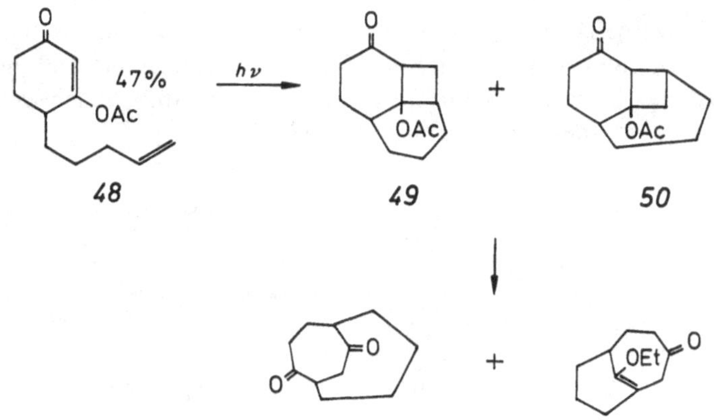

48 *49* *50* *51* *52*

The key step in Büchi's total synthesis of loganin (*53*) [23] made use of the expansion of a four-membered ring to an oxygen-containing heterocycle by retro-aldol reaction. This strategy was later exploited by Tietze in his total synthesis of hydroxyloganin (*54*) as well as hydroxyloganic acid (*55*) [24].

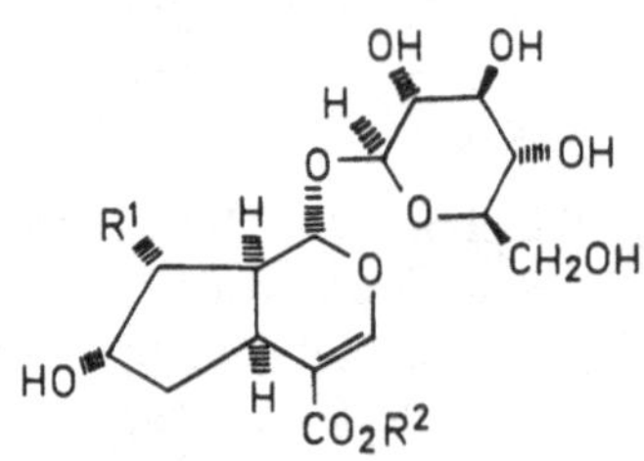

53 R^1 = Me, R^2 = Me *54* R^1 = CH_2OH, R^2 = Me *55* R^1 = CH_2OH, R^2 = H

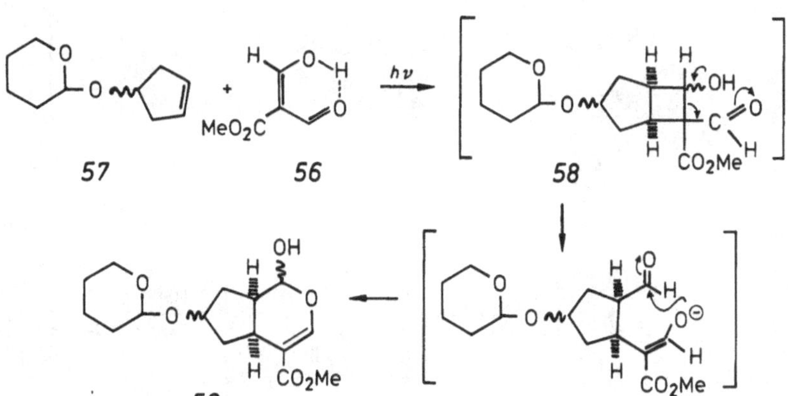

57 *56* *58* *59*

A solution of the ester (*56*) and the tetrahydropyranyl ether (*57*) was irradiated to form the intermediate compound (*58*), which would rearrange through a retro-aldol reaction and a hemiacetal formation route to the less strained six-membered heterocycle (*59*). The hemiacetal (*59*) could be converted to loganin (*53*) in several steps [23].

The total synthesis of optically active loganin (*53*) was accomplished by Partridge [25], who photolyzed a solution of the chiral acetate (*60*) and the ester (*56*). The intermediate (*61*) rearranged to the chiral hemiacetal (*62*), which served as a key compound in the synthesis of chiral loganin (*53*) [25].

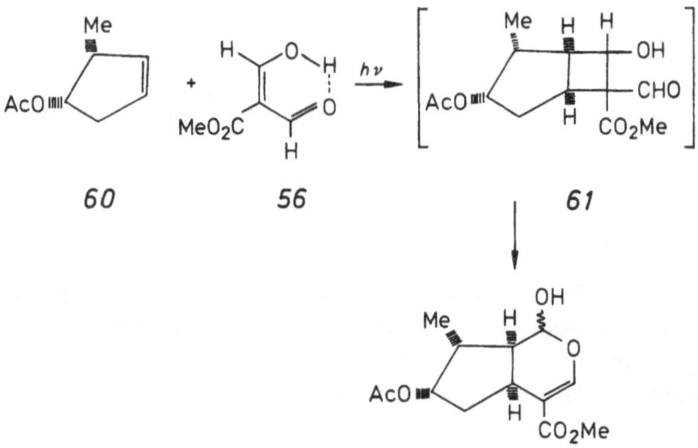

60 56 61

62

A recent paper [26] also reported that the photo [2 + 2] reaction between the olefin (*63*) and the ester (*56*) would yield the desired photoadduct (*64*), which was then reduced by $LiAlH_4$ to afford the diol (*65*). The diol (*65*), upon oxidation, gave the α-methylene-δ-lactone (*67*) via the intermediate lactol (*66*) (Table 3) [26].

The Lewis acid catalyzed thermal [2 + 2] cycloaddition between 1-cyanocyclopentene (*68*) and 1-(N,N-diethylamino)-1-propyne (*69*) afforded the enamine

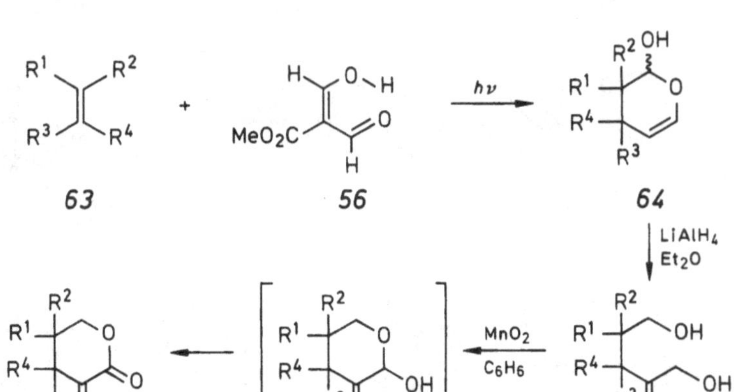

63 56 64

67 66 65

Table 3

R^1	R^2	R^3	R^4	Yield (%) of	
				(*64*)	(*67*)
Me	Me	Me	Me	89	79–80
Me	Me	H	H	85	60–85
H	$-(CH_2)_4-$		H	91	55–70

(*70*) [27], which was hydrolyzed stereoselectively to the cyclobutanone (*71*). The cyclobutanone (*71*) could be converted stereoselectively to the two diastereomeric δ-lactones (*74*) and (*79*) by the following routes [27].

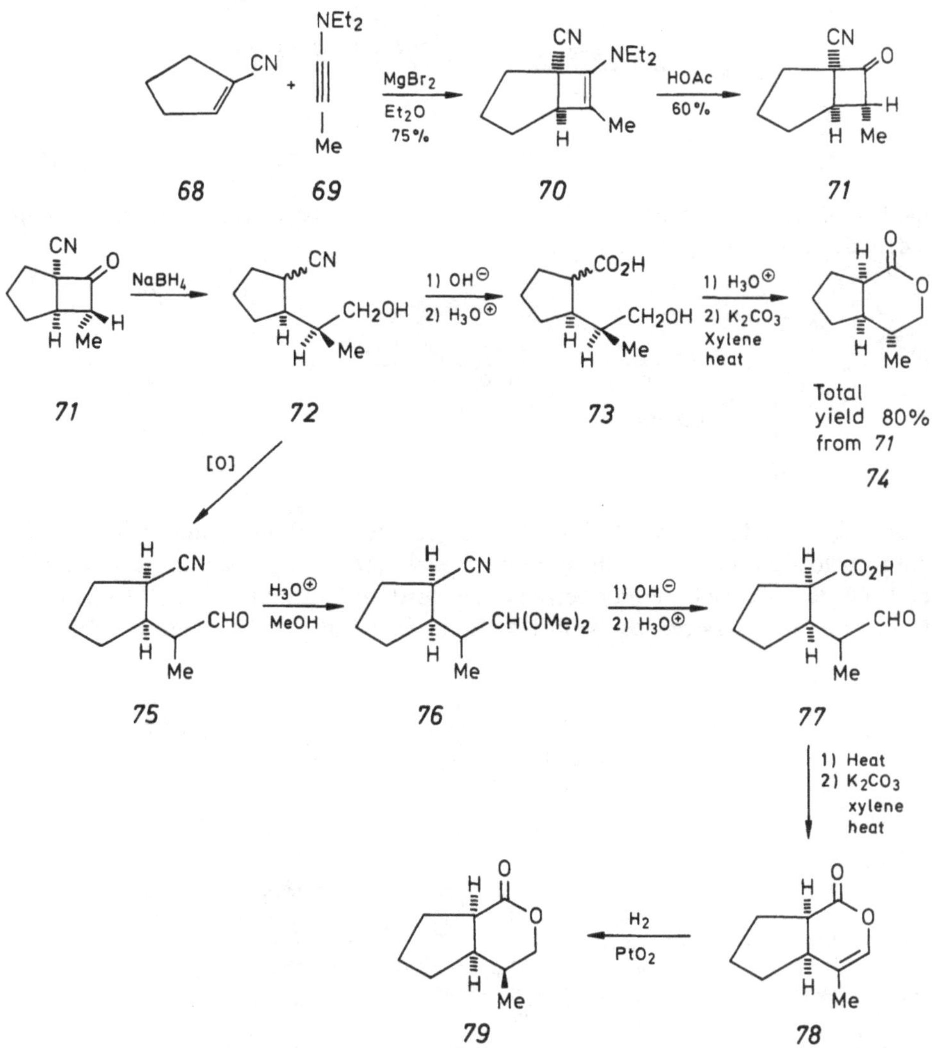

The photo [2 + 2] reaction between 5-methyl-2,3-dihydro-3-furanone (*80*) and methyl β-acetoxyacrylate (*81*) afforded a mixture of four adducts, from which the lactone (*82*) was isolated. The lactone (*82*) was reduced to the alcohol (*83*), which would rearrange quantitatively to the lactone aldehyde (*84*) upon treatment with

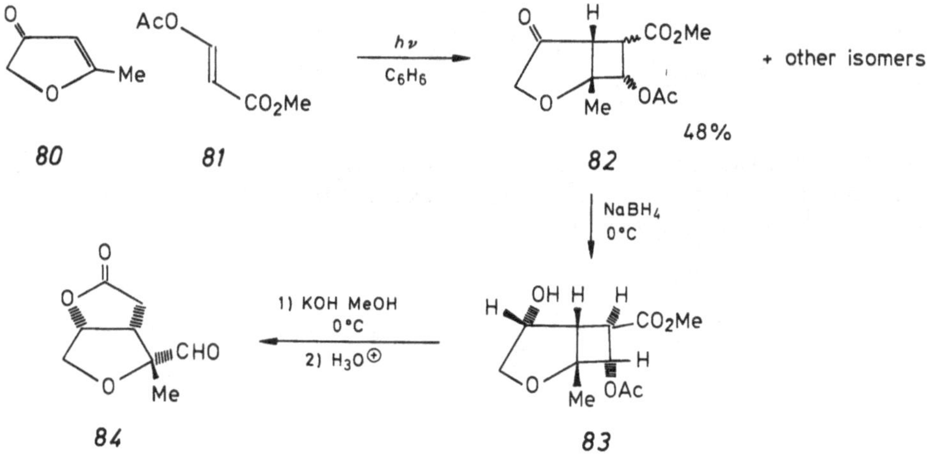

methanolic KOH [28]. Similarly, 2-cyclopentenone can be converted to the lactone aldehyde (*85*) by using the same procedure [28].

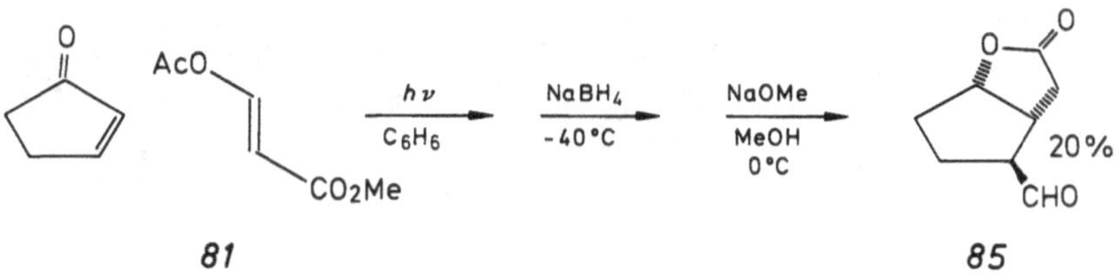

A slight modification of this method was reported by Vandewalle [29]. The photocycloaddition between the cyclopentenones (*86*) and β,β-diethoxyacrylate (*87*) at −40 °C afforded stereoselectively the head to tail adduct (*88*) (Table 4) [29]. Compound (*88*) was reduced with $NaBH_4$ in EtOH at −40 °C, and was followed

Table 4

	R¹	R²	Temp. (°C)	Yield (%)	
				(*88*)	(*89*)
(*86a*)	H	H	25	82.5	59 [overall from (*86a*)]
			−40	94	
(*86b*)	Me	H	25	83.5	53
			−40	91.5	
(*86c*)	H	Me	25	71	45
			−40	100	

by $TsOH \cdot H_2O$ initiated ring opening and lactonization to yield the lactone ester (*89*) (Table 4) [29].

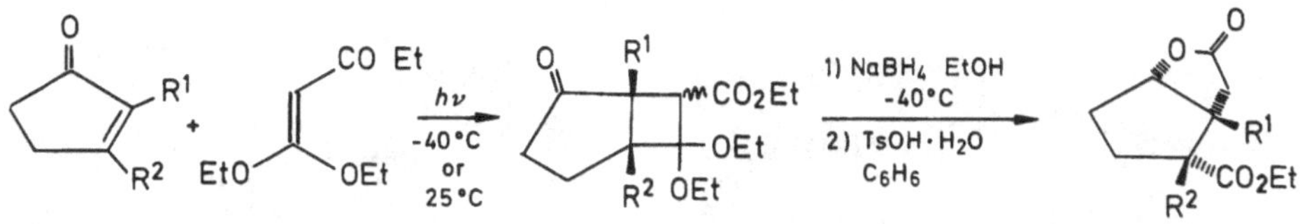

86 87 88 89

Photo-addition of allene to the enone (*90*) yield adduct (*91*) in 75% yield, which was subjected to ketalization in 77% yield. Epoxidation of (*92*) with perbenzoic acid followed by chromatography on alumina afforded two expoxides (*93*) and (*94*). Both (*93*) and (*94*) could be converted separately through (*95*) and (*96*) respectively which was the common intermediate leading to isoishwarane (*98*) and ishwarane following a deketalization-retroaldol-aldol process to furnish the keto-alcohol (*97*) (*99*) [30].

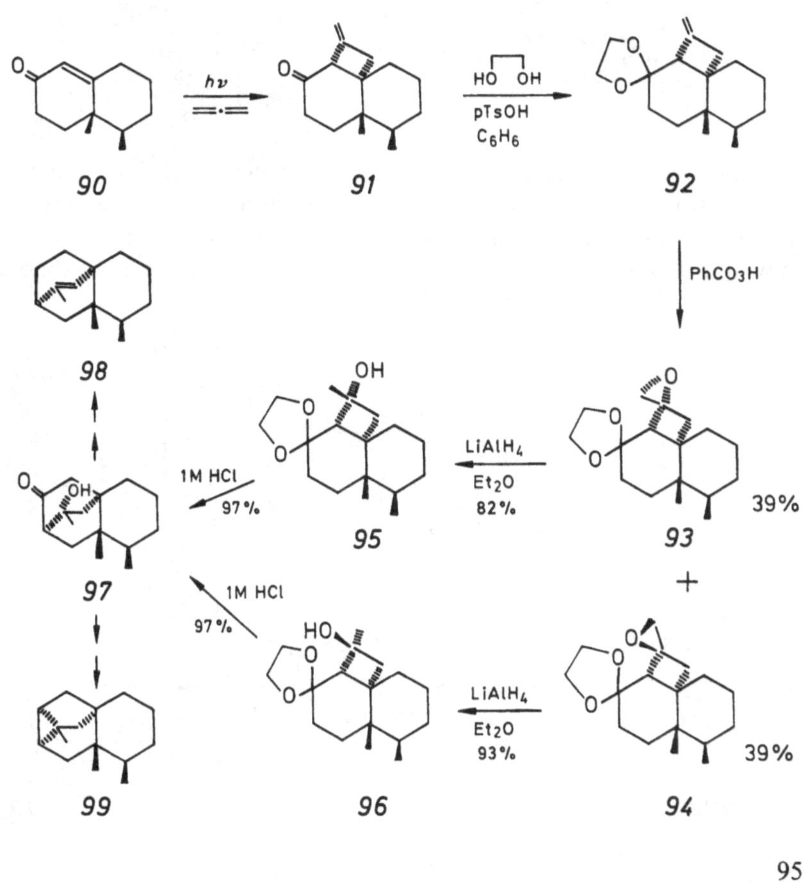

90 91 92 93 94 95 96 97 98 99

The deketalization-retroaldol-aldol procedure was applied to compound (*100*). Acid catalyzed deketalization, followed by the retroaldol-aldol process afforded (*101*), which could be transformed to the diterpene trachylobane (*102*) [31].

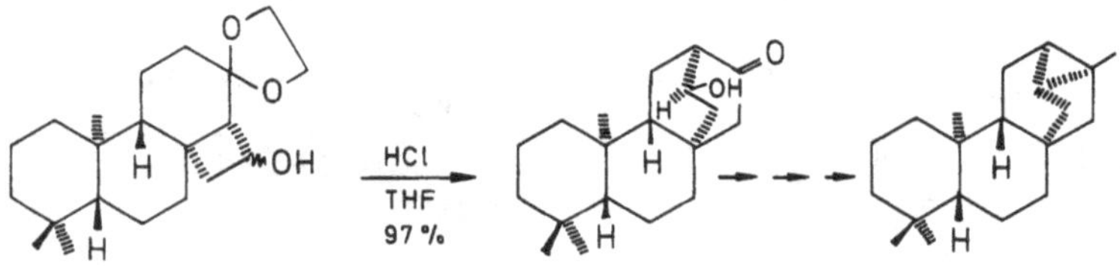

100 *101* *102*

The key steps in the synthesis of the stemodane-type diterpenoids are again the retroaldol-aldol procedure. This was best demonstrated by the rearrangement of the ketal (*103*) to the epimeric alcohol (*104*), upon treatment with acid, in 60% yield [32].

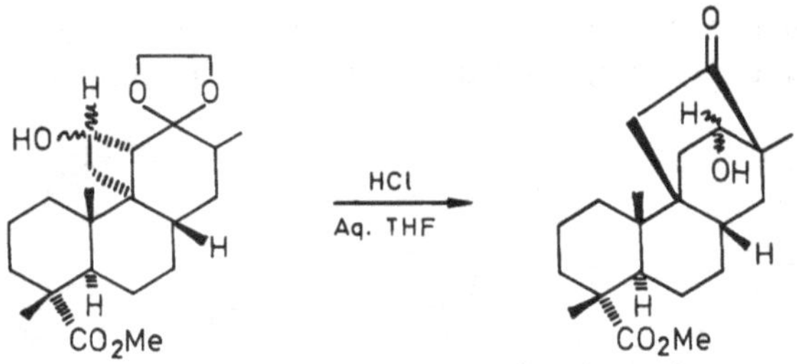

103 *104*

Treatment of the ketone (*105*) with methanolic KOH led to hydrolysis of the acetate group, retro-aldol reaction and aldol cyclization, which eventually afforded a mixture of epimeric alcohols (*106*) [33]. This strategy was used to construct the basic skeletons of various alkaloids, e.g. talatisamine [34], atisine [35] and chasmanine [36].

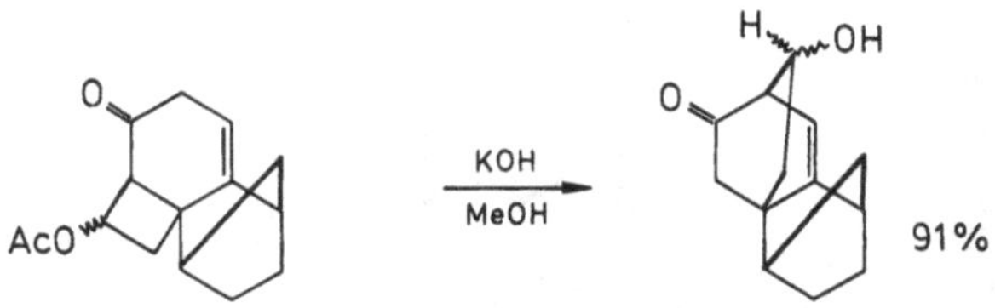

105 *106*

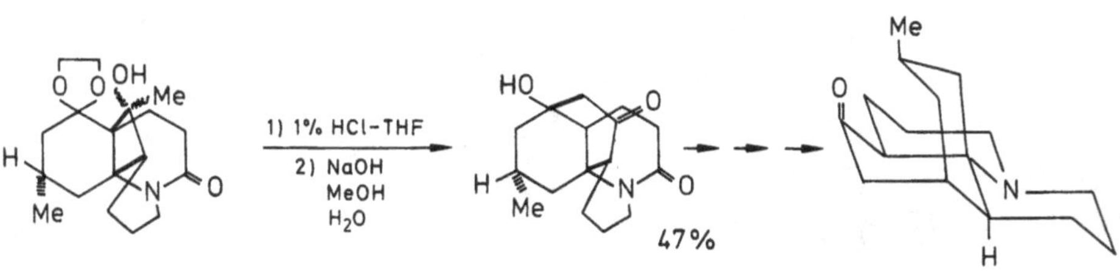

107 *108* *109*

The empimeric alcohol (*107*) again underwent the deprotection-retroaldol-aldol process to afford the product (*108*), which was used as a key intermediate in the total synthesis of 12-epi-lycopodine (*109*) [37].

The 1,5-diketones isolated from the de Mayo reaction can usually be subjected to aldolization process, which yields cyclohexenones. Thus, the de Mayo reaction can be regarded as an useful annelation process. Although in principle, the orientation of the [2 + 2]cycloaddition of an alkene to a β-diketone can be controlled by solvent polarity [7], in practice, the stereochemistry of the [2 + 2]cycloaddition between an unsymmetrical alkene and an unsymmetrical β-diketone can hardly be controlled. The resulting 1,5-diketones would yield eight strucurally different cyclohexenones through an aldol reaction, not counting stereoisomers. This disadvantage often restricts the practical applicability of the de Mayo reaction. However, the use of α-formyl ketones [8], β-acetoxy-α,β-unsaturated esters [9, 28] and β-trimethylsiloxy-α,β-unsaturated esters [13] somehow compensates this discrepancy, and usually provides pure 1,5-diketones. In any case, if the stereochemistry of the [2 + 2]cycloaddition can be controlled, the main obstacle for the extensive application of the de Mayo reaction will be removed.

2.2 Fragmentation Reaction

Fragmentation [38] occurs on compounds of the form W—C—C—X, where X is a good leaving group and W in most cases is $RO{-}\overset{|}{\underset{|}{C}}{-}$ (Eq. (1)).

$$R{-}\ddot{O}{-}\overset{|}{\underset{|}{C}}{-}\overset{|}{\underset{|}{C}}{-}\overset{|}{\underset{|}{C}}{-}X \rightarrow R{-}\overset{+}{O}{=}\overset{|}{\underset{|}{C}} + \overset{|}{\underset{|}{C}}{=}\overset{|}{\underset{|}{C}} + X^{-} \qquad \text{(Eq. 1)}$$

The mechanisms of this fragmentation reaction are believed to be mostly E1 or E2. When W and X groups are appropiately situated on a four-membered ring, the strain of the four-membered ring would assist the fragmentation so that the reaction would be even more facile and as a result, larger ring system can result. The construction of the four-membered carbocycles will again make use of the photo[2 + 2]cycloaddition. As we have already seen, photo[2 + 2]cycloadditions usually lack stereo-control. However, this small shortcoming does not devaluate the importance of fragmentation reactions of cyclobutane derivatives, especially when this disadvantage is overcome by using intramolecular [2 + 2]photocyclizations. The following reactions are some typical examples of these transformations. Extremely complex cyclic systems can normally be obtained through the use of fragmentation processes. The fragmentation reactions of four-membered carbocycles

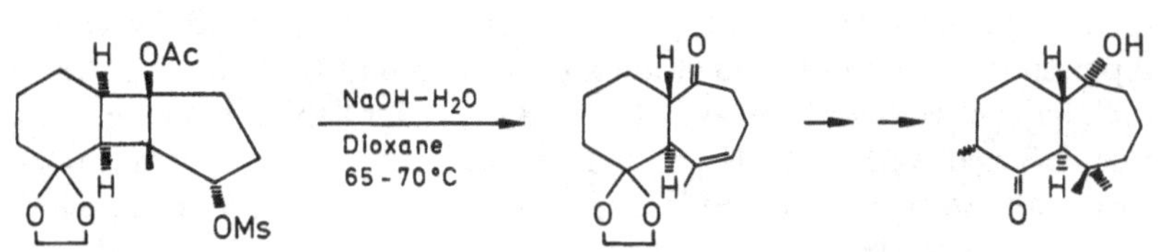

110 *111* *112*

are generally used to construct larger ring compounds. A recent review by Oppolzer [39] has given detailed descriptions of this field.

The fragmentation reaction was applied most successfully in the total synthesis of sesquiterpenes [39]. For example, the mesylate (*110*) underwent fragmentation reaction to provide the ketone (*111*), which could be converted to (*112*). The compound (*112*) was eventually transformed to the isolable mixtures of β-himachalene (*113*), *trans*-α-himachalene (*114*) as well as *trans*-γ-himachalene (*115*) [40].

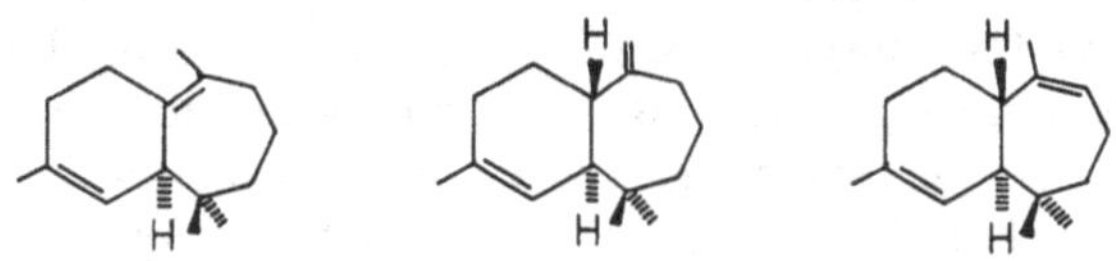

113 *114* *115*

Photocycloaddition of cyclopentene (*116*) to the trimethylsilyl ether (*117*) gave 2 stereoisomers (*118*) and (*119*). When (*118*) was reduced with $LiAlH_4$ in Et_2O, followed by treatment will MsCl, pyridine and dimethylaminopyridine (DMAP), the silyl ether (*120*) was obtained. Finally, (*120*) was allowed ro react with KF in the presence of [18]-crown-6 in CH_2Cl_2, as a result, hexahydroazulenone (*121*) was isolated [41]. On the other hand, the cyclobutane (*119*) could also be converted similarly to (*121*) in an overall yield of 47% [41].

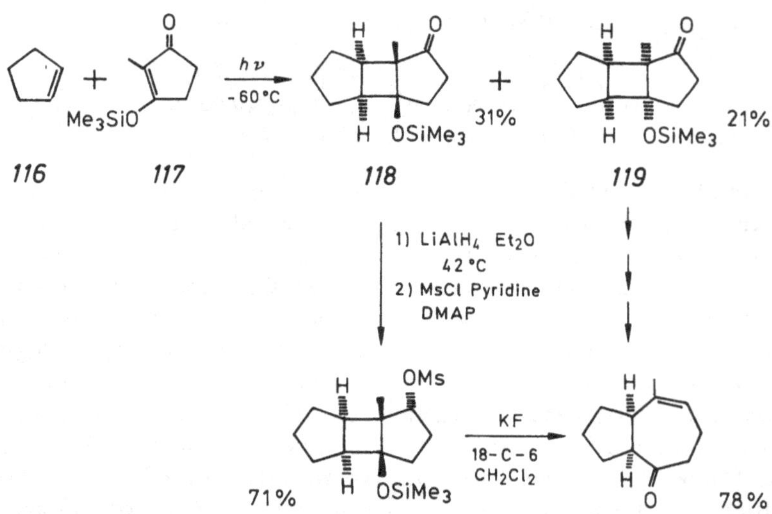

The chemistry of cyclobutanes was remarkably exploited by Oppolzer [42], who reported that the aldol condensation of the acetal (*122*) with 1,2-*bis*(trimethylsiloxy)-cyclobutene was catalyzed by $BF_3 \cdot Et_2O$ to afford the cyclobutanone (*123*), which was refluxed with pTsOH in benzene to give 1,3-cyclopentanedione (*124*). On treatment with acetylchloride in pyridine at 0 °C, the compound (*124*) was transformed to the enol acetate (*125*). Irradiation of (*125*) gave the intramolecular photoadduct (*126*). Then on successive treatments with MeMgI, KOH and MsCl in

Et_3N, the adduct (*126*) yielded azulenone (*127*). The ketone (*127*) reacted with Wittig reagent to afford a [3.3:1] mixture of 1-epi-β-bulnesene (*128a*) and β-bulnesene (*128b*) [42].

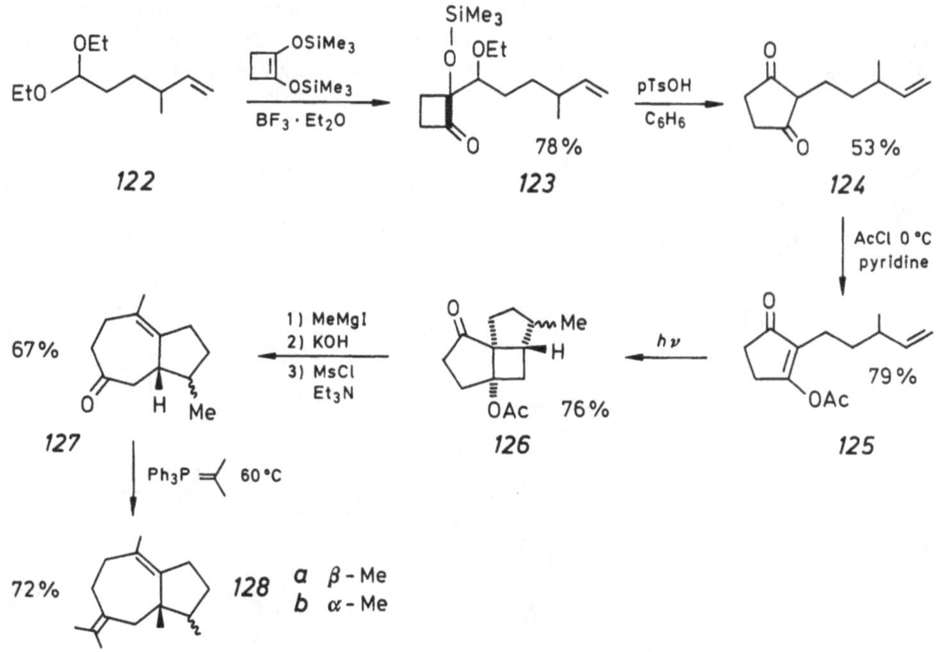

When the monotosylate (*129*) was treated with K_2CO_3 in 60% aq. acetone in a sealed tube at 85 °C for 1 day, the compound (*130*) was obtained as a single product. Compound (*130*) could be converted to (±)-hirsutene (*131a*) by further reactions [43]. Furthermore, (*130*) could be also transformed to (±)-coriolin (*131*b) [44].

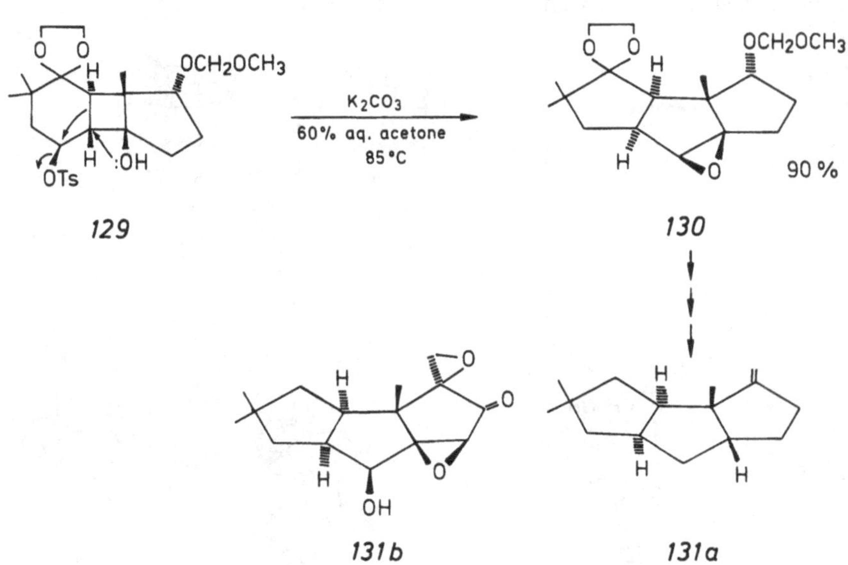

A new formal synthesis of (±)-zizaene (*136*) via the fragmentation reaction is a typical example[45,46]. When the enol acetate (*132*) was irradiated, the isolable photoadduct (*133*) was formed together with other isomers. Reduction and mesylation of which afforded the mesylate (*134*). Then, fragmentation reaction would yield an epimeric mixture of the alkenone (*135*) in the presence of NaOH in aqueous dioxane at 60 °C, thus formulated a total synthesis of (±)-zizaene (*136*)[45,46].

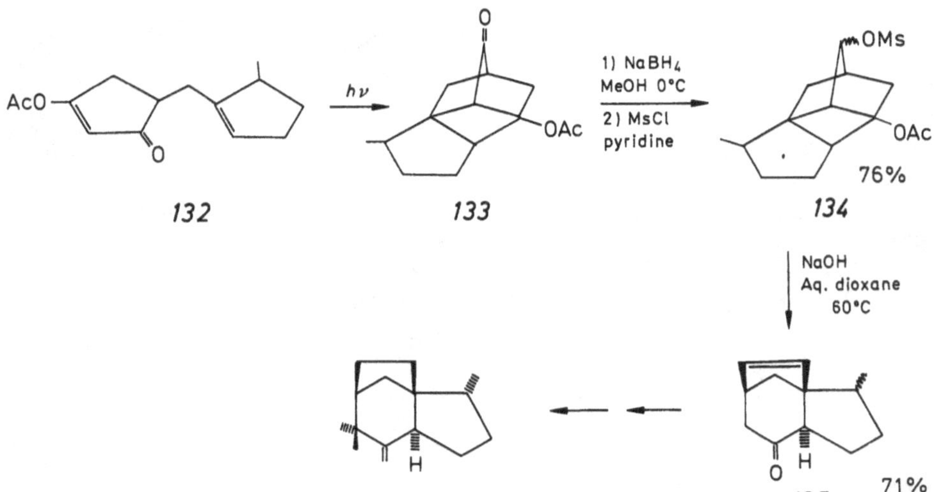

132 *133* *134* *135* *136*

The enone (*137*) was allowed to react with MeMgBr—Cu_2I_2 complex at 0 °C to furnish the keto-ester (*138*), which was then reduced by $NaBH_4$ to afford the lactone (*139*). On the other hand, consecutive thioketalization, desulfurization and reduction of the ester (*138*) gave the alcohol (*140*), which was allowed to react with *p*TsCl in pyridine to afford (*141*) after smooth and concomitant fragmentation. The compound (*141*) was converted to 5-epi-kessane (*142*) and dehydrokessane (*143*)[47].

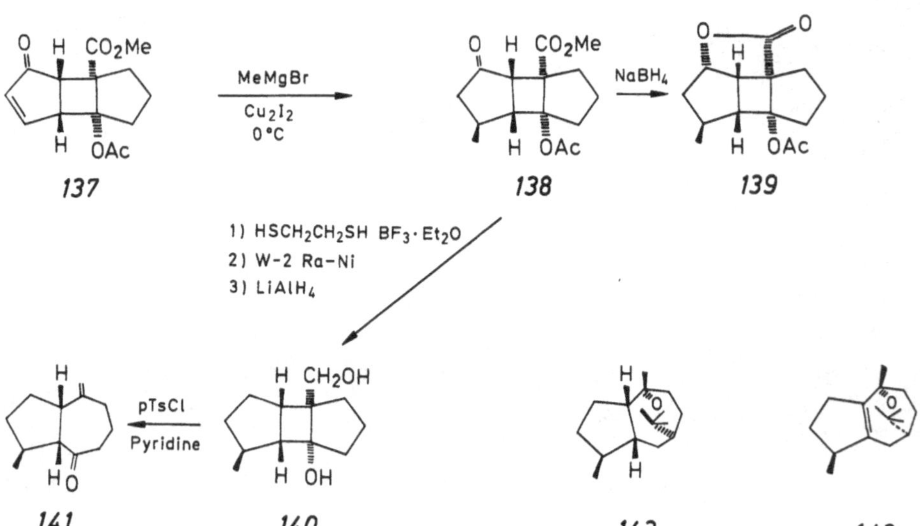

137 *138* *139* *141* *140* *142* *143*

Treatment of (*144*) with Na_2CO_3 in aqueous MeOH gave the ketoaldehyde (*145*) [48], which possessed the *Ormosia* alkaloid skeleton.

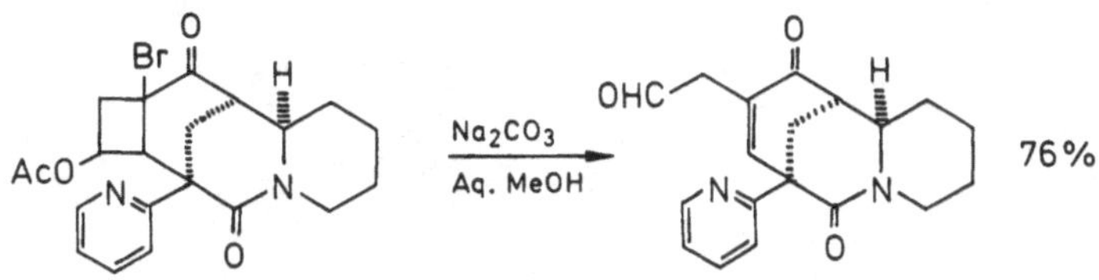

144 *145*

2.3 Ring Expansion to 5-Membered Carbocycles and Heterocycles

The ring expansion of cyclobutane derivatives to other carbocycles remains to be one of the most powerful tool in synthetic organic chemistry. Cyclobutanones are exceptionally facile starting materials for the preparations of γ-lactones as well as cyclopentanones.

The preparation of cyclobutanones can easily be realized by using the thermal-[2 + 2]cycloaddition of haloketenes to olefinic compounds (Eq. (2)) [49].

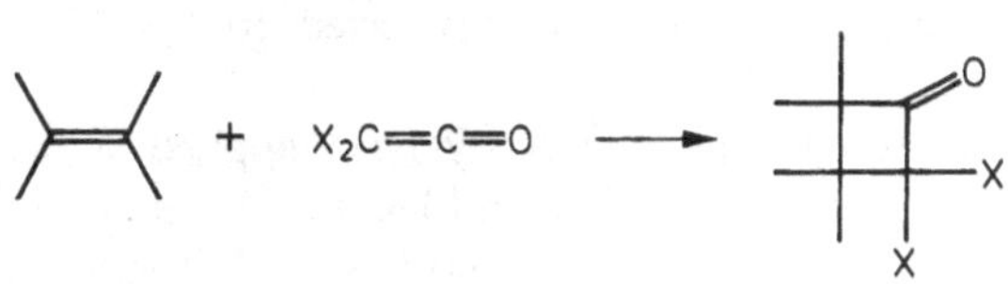

(Eq. 2)

$$CHX_2-\overset{O}{\overset{\|}{C}}-X \xrightarrow[\text{hexane}]{Et_3N} X_2C{=}C{=}O + Et_3\overset{\oplus}{N}HCl^{\ominus} \qquad \text{(Eq. 3)}$$

$$CX_3-\overset{O}{\overset{\|}{C}}-X \xrightarrow[Et_2O]{Zn(Cu)} X_2C{=}C{=}O + ZnX_2 \qquad \text{(Eq. 4)}$$

Consequently, haloketenes can be readily generated *in situ* by two most widely used methods (a) the triethylamine dehydrohalogenation of an acyl halide (Eq. (3)) [50] (b) the dehalogenation of an α-haloacyl halide with activated zin (Eq. (4)) [51]. Since the halogen substituents on the cyclobutanone can be reductively removed by usual procedures, the synthesis of a halocyclobutanone constitutes a formal preparation of the cyclobutanone, the synthetic utility of which is convincingly demonstrated by the following examples.

Replacement of one of the chloro substituents on the ketone (*146*) with Me_2CuLi in the presence of MeI, followed by ring expansion and concomitant esterification with CH_2N_2 provided the 5-membered ketone (*147*) [52].

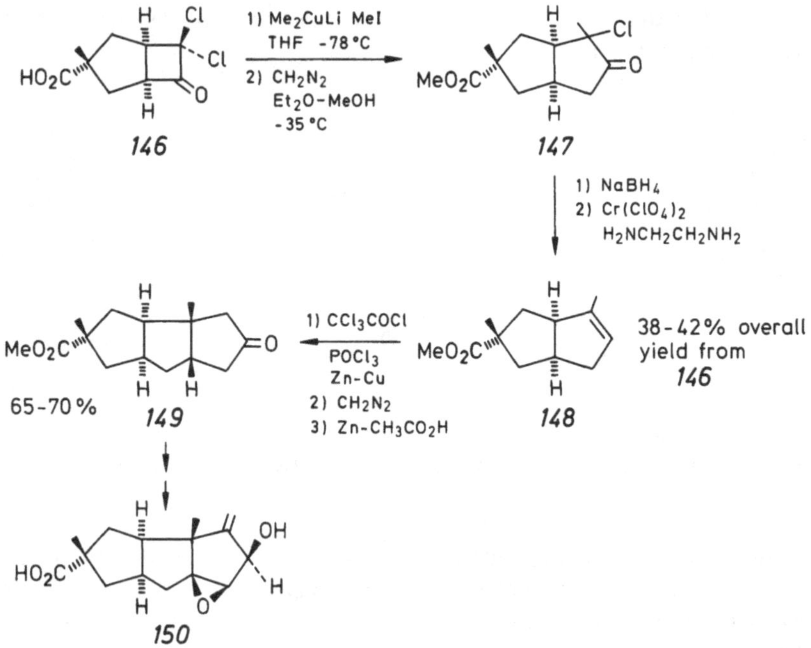

Reduction of the ketone (*147*), followed by elimination gave the olefin (*148*). The olefin (*148*) was subjected again to a second annelation, and as expected, dichloroketene addition, ring expansion and zinc reduction gave the tricyclic compound (*149*). Compound (*149*) could be converted to (±)-hirsutic acid C (*150*) [52].

The diazomethane methodology has previously been applied by several other groups to expand cyclobutanones. For example, in Fleming's loganin synthesis, [53, 54] when the cyclobutanone (*151*) was allowed to react with diazoethane, (*152*) was isolated as a mixture of stereoisomers. Removal of the chlorine atom and equilibration of the methyl group with NaOMe converted (*152*) to the ketone (*153*) [53, 54].

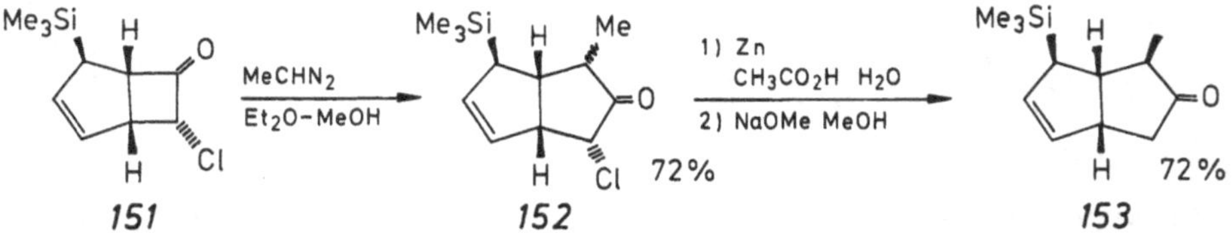

Table 5

R¹	R²	R³	R⁴	R⁵	Yield of (*155*)
Me	H	H	Me	H	89%
H	H	H	H	H	84%
H	Me	H	Me	H	83%
H	H	Me	Me	H	92%

In a more practical sence, treatment of ethyl diazoacetate and boron trifluoride etherate transformed the cyclobutanone (*154*) to (*155*) (Table 5) [55].

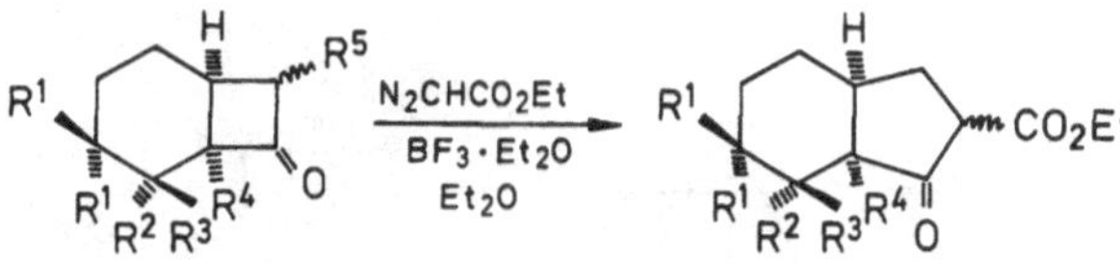

154 *155*

On the other hand, the regiospecificity has totally changed in the case of (*156*), which provided (*157*) in quantitative yield [55]. Whereas the reaction of the ketone (*158*) showed very weak regioselectivity, from which (*159*) and (*160*) were separated [55].

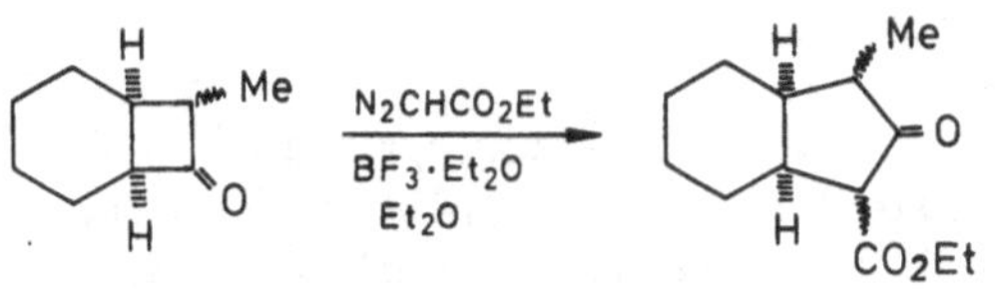

156 *157*

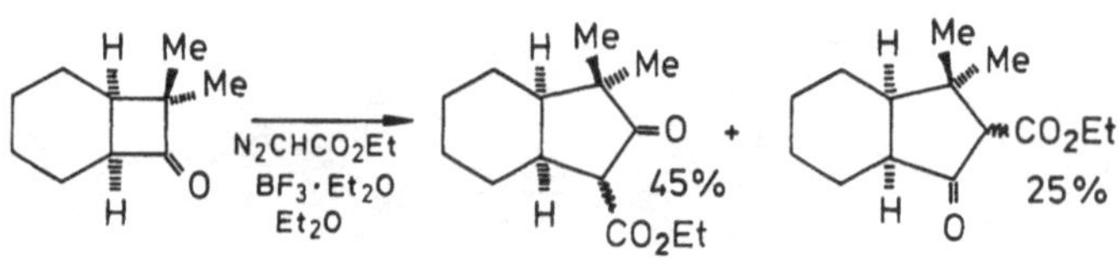

158 *159* *160*

In Paquette's total synthesis of ($\pm$)-pentalenene, [56] the key step involved the ring expansion of (*162*) to (*163*) by treatment with CH_2N_2 [56].

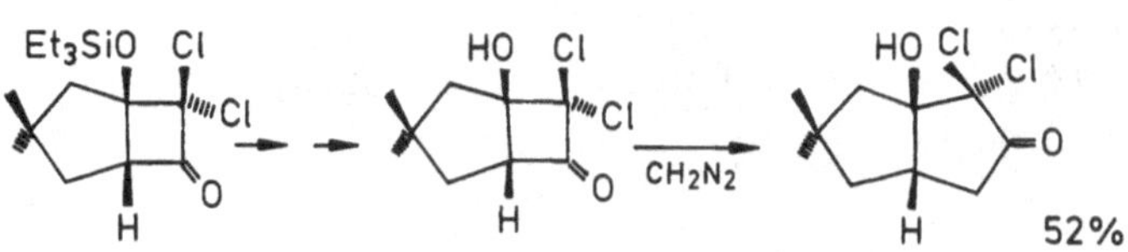

161 *162* *163*

The condensation of cyclopropylide with carbonyl compounds furnishes oxaspiropentane in excellent yields [57]. The presence of an oxygen atom plus the inherent high strain energy of the oxaspqropentane renders it extremely reactive, and hence causes facile rearrangement to a cyclobutanone upon treatment with aq. HBF_4, $LiBF_4$ or $LiClO_4$ (Eq. (5)) [57]. Such strategy formulates an important method of making cyclobutanones from carbonyls. Subsequent treatment of the cyclobutanone with sulfur ylide yields oxaspirohexane, which in turn rearranges to a cyclopentanone on treatment with LiBr (Eq. (6)).

The approach to cyclopentanone derivatives from oxaspirohexanes can be appropiately demonstrated by the following examples [57].

Trost's approach to gibberellins applied the ring expansion of the epoxides (*165*) and (*168*), which would provide the spirocyclopentanone (*166*) [58]. The cyclobutanone (*164*) could be converted to (*165*) by reaction with dimethylsulfonium methylide in THF and then LiBr in ØH-HMPA effected the rearrangement of (*165*)

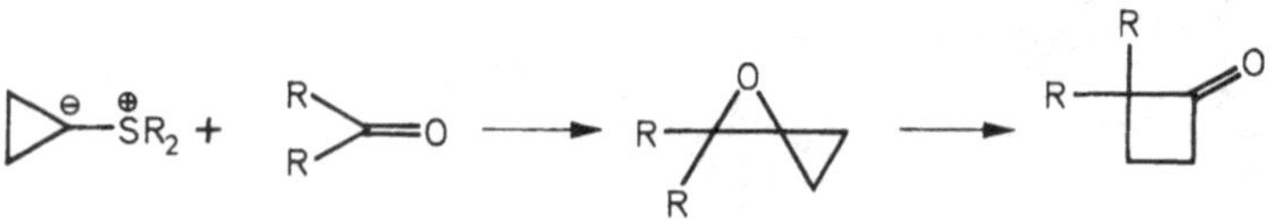

(Eq. 5)

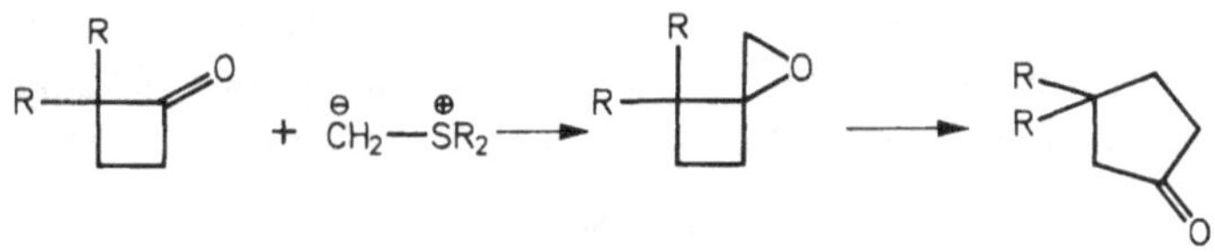

(Eq. 6)

to (*166*) [58]. The driving force of this reaction is again attributable to the strain energy of the oxaspirohexane (*165*). Alternatively, (*164*) could be converted to (*167*) by Wittig olefination. Epoxidation followed by LiBr promoted rearrangement afforded (*166*) in an overall yield of 78% from (*164*) [58].

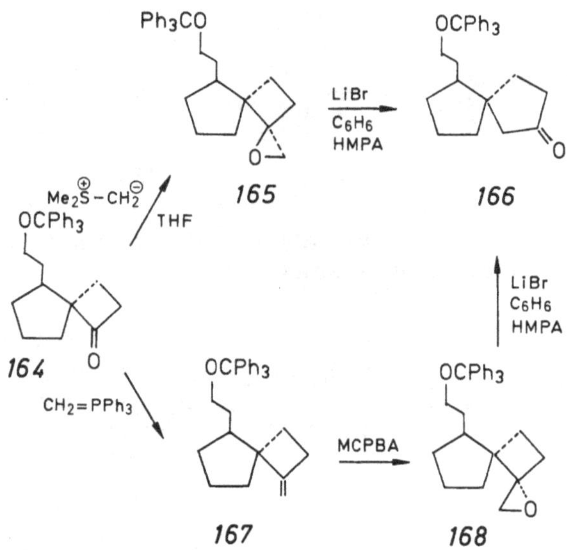

Table 6

R^1	R^2	R^3	R^4	Yield of (*170*)
H	$C_{10}H_{21}$	H	H	91%
Me	C_9H_{19}	H	H	77%
H	$C_{10}H_{21}$	H	C_5H_{11}	87%
H	$C_{10}H_{21}$	H	H	91%
H	H	Me	C_9H_{19}	83%

Lithium iodide in refluxing CH_2Cl_2 has also been used to convert the oxaspirohexanes (*169*) to polyalkylated cyclopentanones (*170*) (Table 6) [59].

The β-hydroxyselenide (*172*) was prepared by reaction of the cyclobutanone (*171*) and 2-lithio-2-methylselenopropane. The rearrangement of (*172*) to α-cuparenone (*173*) could be promoted by a few conditions (Table 7) [60]. On the other hand, (*174*)

Table 7

	condition	yield
(*172*)→(*173*)	$AgBF_4/Al_2O_3/CH_2Cl_2$/20 °C	69%
	$TlOEt/CHCl_3$/20 °C	57%
	CH_3OSO_2F/Et_2O/20 °C	82%
(*176*)→(*177*)	LiI/CH_2Cl_2/40 °C	80%
	LiBr/ØH/HMPT/80 °C	80%
	LiI/dioxane/12-crown-4/80 °C	94%

could be converted to (*175*) by treatment with methylselenomethyllithium, but on reaction with thallium ethoxide in $CHCl_3$, (*175*) was converted only to the oxaspirohexane (*176*), which was converted to β-cuparenone (*177*) under several conditions (Table 7) [60].

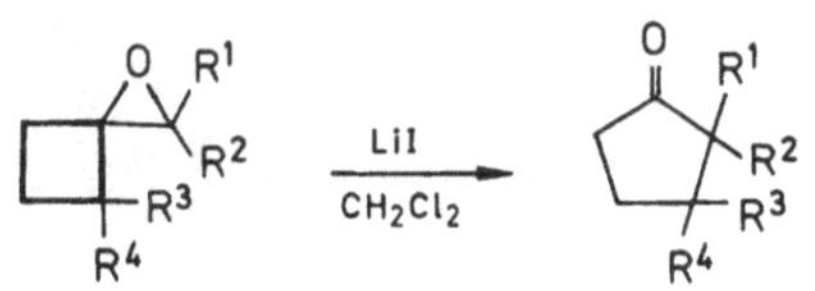

169 170

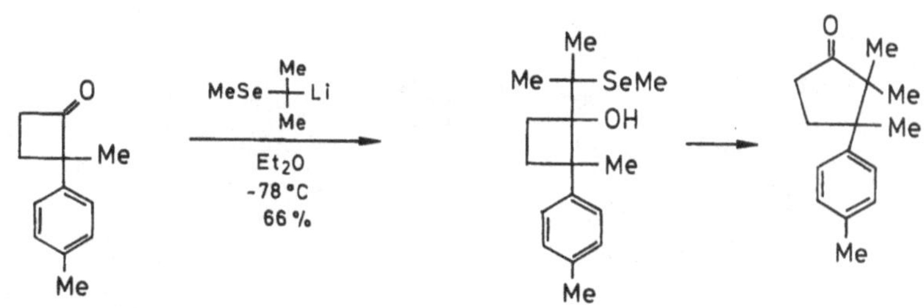

171 172 173

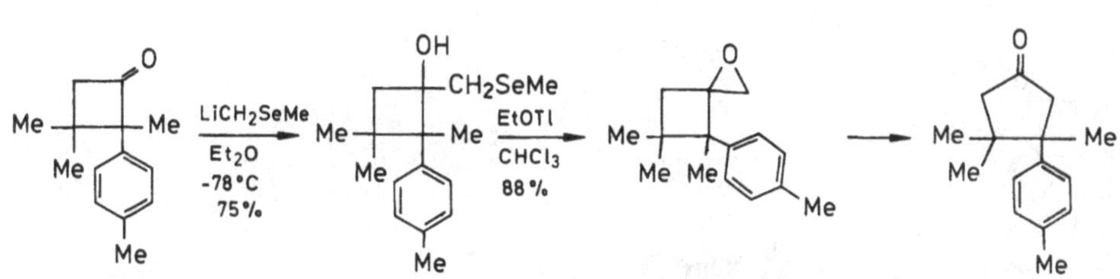

174 175 176 177

α-Lithioselenoxides (*179*) reacted with the cyclobutanone (*178*) smoothly and the pinacol-like rearrangement of the initial product (*180*) led to the cyclopentanones (*181*) (Table 8) [61].

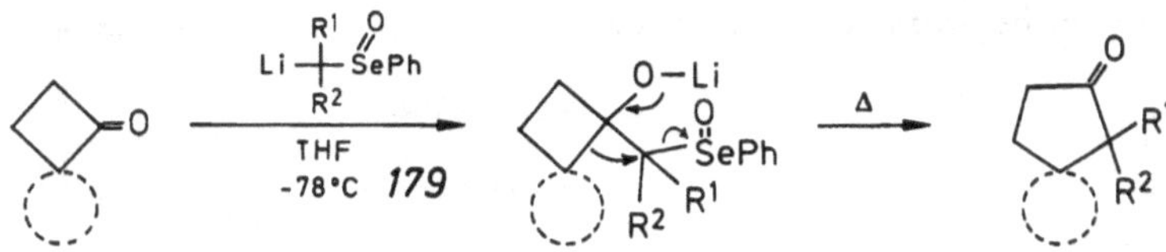

178 180 181

Table 8

(*178*)	(*179*)	(*181*)	Yield of (*181*)
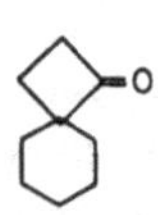	R^1 = H, R^2 = H	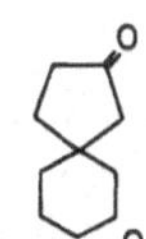	82%
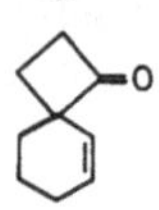	R^1 = H, R^2 = H	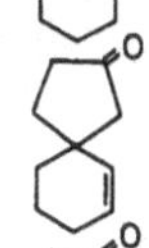	73%
"	R^1 = Me, R^2 = H	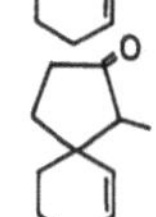	71%
"	R^1 = Me, R^2 = Me	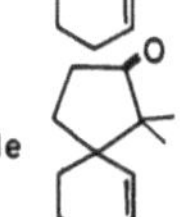	63%
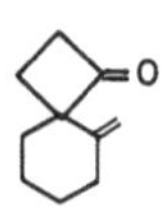	R^1 = H, R^2 = H	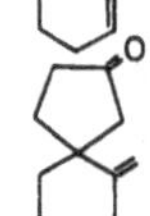	88%
"	R^1 = Me, R^2 = H	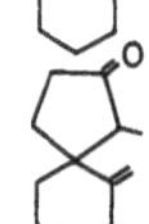	93%
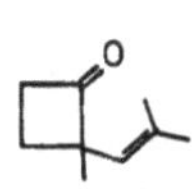	R^1 = H, R^2 = H	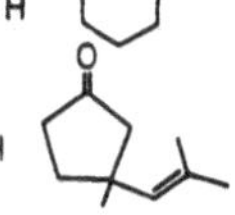	82%
"	R^1 = Me, R^2 = Me	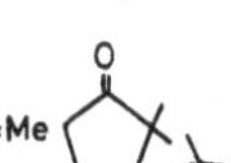	52%
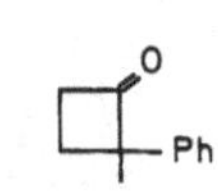	R^1 = H, R^2 = H	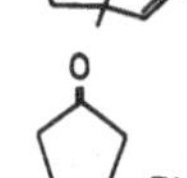	92%
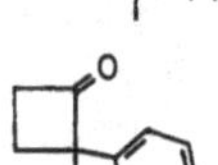	R^1 = Me, R^2 = Me	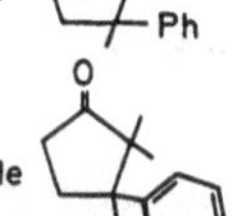	39%

173

1,2-Bis(trimethylsiloxy)cyclobutene, a very useful building block, is easily prepared by the modified acyloin condensation of diethyl succinate with sodium in xylene in the presence of trimethylsilylchloride [62].

1,2-Bis(trimethylsiloxy)cyclobutene reacted with the aldehyde (*182a*) or the acetal (ketal) (*182b*) to give the pinacol (*183a*) or (*183*b) through either a Lewis acid mediated aldol addition or a boron trifluoride etherate catalyzed addition [63,64]. Treatment of the pinacol (*183a*) or (*183b*) with trifluoroacetic acid led to the isolation of 1,3-cyclopentanedione (*184a*) or (*184b*) (Table 9 and 10) [63].

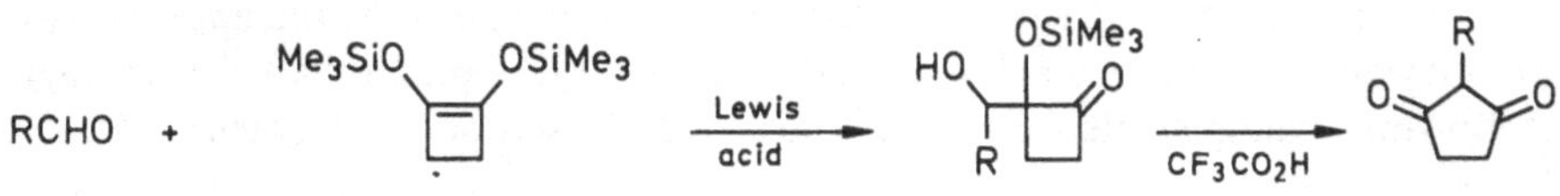

182a *183a* *184a*

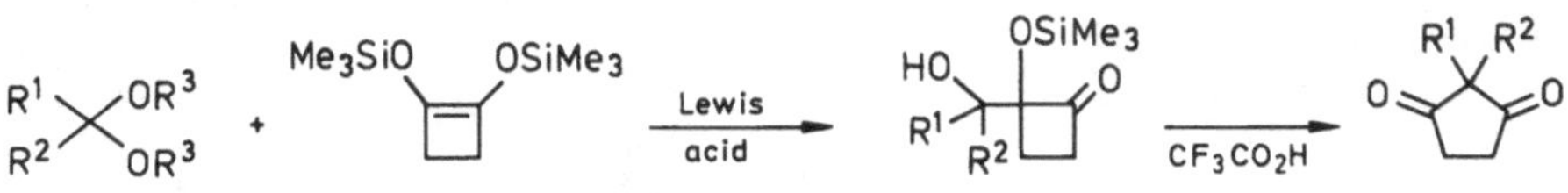

182b *183b* *184b*

Table 9

(*182a*)	Lewis acid	yield of (*183a*)	(*184a*)	yield of (*184a*)
R = ∅	$TiCl_4$	78%	R = ∅	97%
R = ∅	$^nBu_4NF^-$	75%	R = ∅	97%

Table 10

(*182b*)	Lewis acid	Yield of (*183b*)	(*184b*)	Yield of (*184b*)
R^1 = Φ, R^2 = H, R^3 = Et	$BF_3 \cdot Et_2O$	94%	R^1 = Φ, R^2 = H	93%
R^1 = $^nC_9H_{19}$, R^2 = H, R^3 = Et	$TiCl_4$	90%	R^1 = $^nC_9H_{19}$, R^2 = H	87%
R^1 = Et, R^2 = Et, R^3 = Me	$BF_3 \cdot Et_2O$	92%	R^1 = Et, R^2 = Et	87%
R^1, R^2 = —$(CH_2)_5$—, R^3 = Et	$BF_3 \cdot Et_2O$	90%	R^1, R^2 = —$(CH_2)_5$—	88%
R^1, R^2 = —$(CH_2)_{11}$—, R^3 = Me	$BF_3 \cdot Et_2O$	92%	R^1, R^2 = —$(CH_2)_{11}$—	94%
R^1, R^2 =  , R^3 = Me	$BF_3 \cdot Et_2O$	60%	R^1, R^2 = 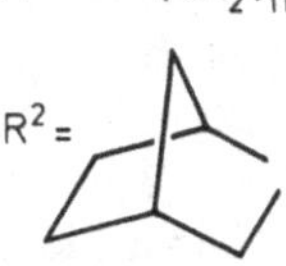	92%

The reaction between ketones and peracids such as perbenzoic acid or peracetic acid, provides esters. (Eq. (7)). The overall result is the "insertion of oxygen". This reaction is known as the Baeyer-Villiger oxidation [65], which is often applied to cyclic ketones to give lactones. Cyclobutanones, upon treatment with peracid, yield

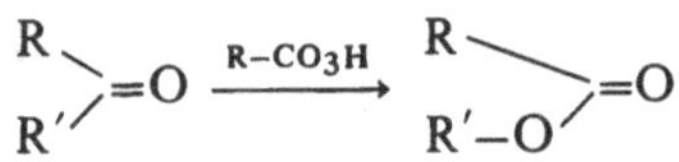
(Eq. 7)

γ-lactones in excellent yields. Since the mechanism of the Baeyer-Villiger oxidation involves the migration of alkyl groups, the order of migration is therefore an important factor in the development of Baeyer-Villiger oxidation for organic synthesis. It has been determined empirically that for unsymmetrical ketones, the approximate order of migration is tertiary alkyl > secondary alkyl, aryl > primary alkyl > methyl. The following examples depict clearly how the "insertion of oxygen" can be controlled.

The allyl silane (*185*) has been used to prepare key intermediates for the synthesis of prostaglandins [66, 54].

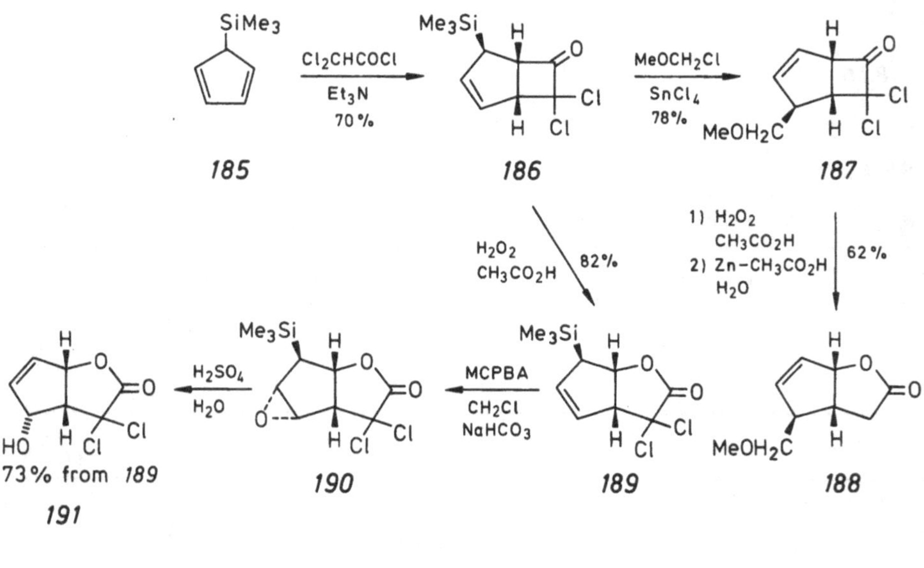

Dichloroketene reacted with (*185*) to form the *gem*-dichlorocyclobutanone (*186*), which could be transformed to the important intermediates (*188*) and (*191*) for the synthesis of prostaglandins A and F series [66, 54].

The key steps in the ring expansion of the cyclobutanones (*186*) and (*187*) are the Baeyer-Villiger oxidation effected by H_2O_2—HOAc. It is noteworthy to point out that the Baeyer-Villiger oxidation is regiospecific and serves to be an excellent method for the preparation of γ-lactone from cyclobutanones.

In Grieco's total synthesis of the antileukemic secoeudesmanolides ivangulin (*194*) [67] and eriolanin (*197*) [68], the Baeyer-Villiger oxidation was again found to be indispensable in the conversion of the cyclobutanones (*192*) and (*195*) to the γ-lactones (*193*) and (*196*) respectively [67, 68].

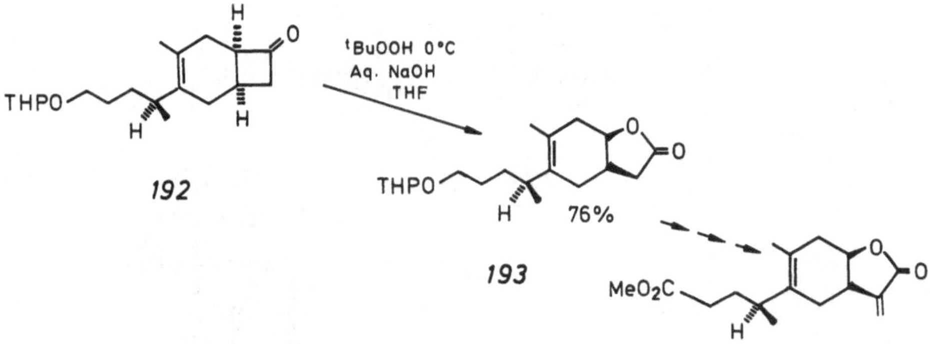

192 *193* *194*

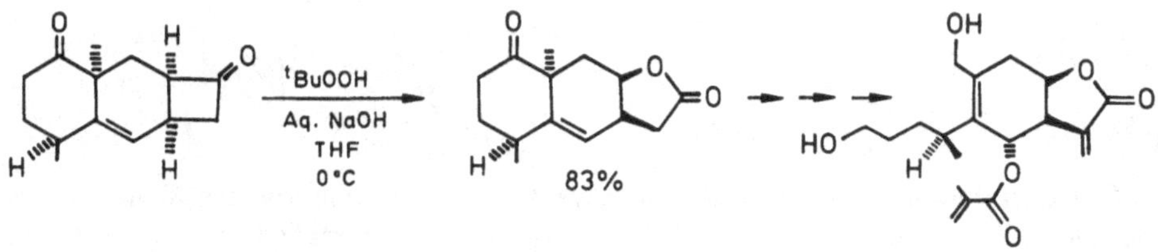

195 *196* *197*

The key step in Smith's paniculides synthesis was the oxidation of the cyclobutanone (*198*) to the epoxylactones (*199*) and (*200*), which were separated by flash chromatography [69].

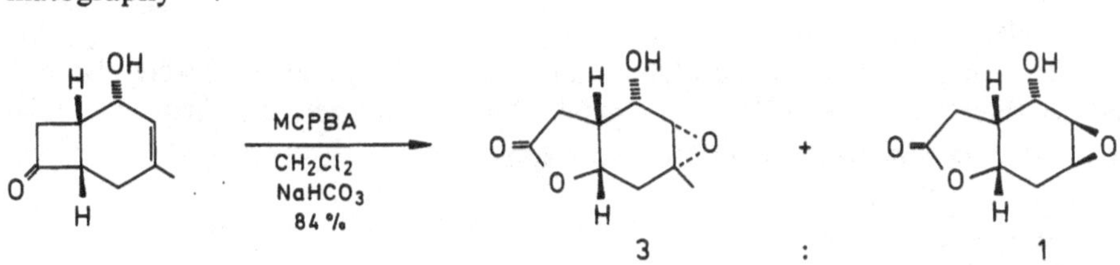

198 *199* *200*

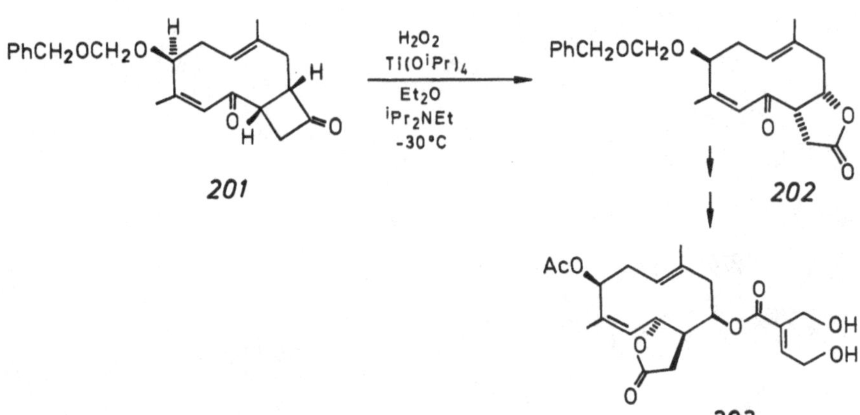

201 *202* *203*

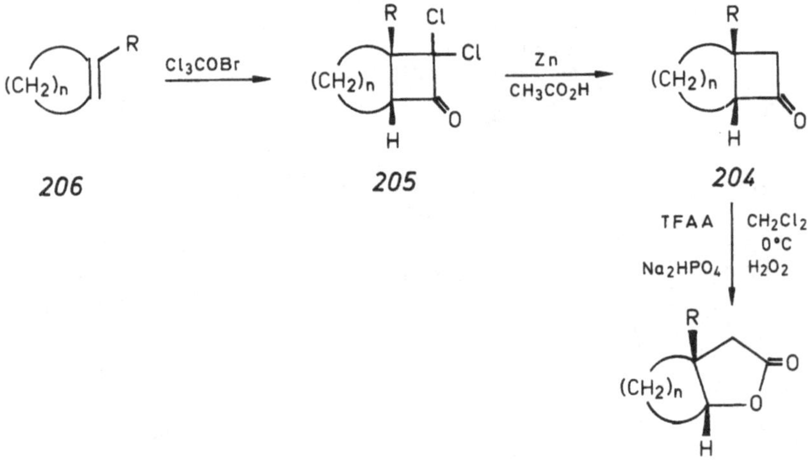

206 205 204 207

The cyclobutanone (*201*) also underwent smooth oxidation to afford the γ-lactone (*202*), which was not isolated and was allowed to react further to yield the cytotoxic germacranolide eucannabinolide (*203*) in a number of steps [70].

A systematic study of the Baeyer-Villiger oxidation of cyclobutanones was recently reported by Jeffs [71]. The cycloalkenes (*206*) reacted readily with dichloro-ketene to give the *gem*-dichlorocyclobutanone (*205*), which were reduced by Zn to the cyclobutanone (*204*). Baeyer-Villiger oxidation of (*204*) yielded the γ-lactone (*207*) in fair yields. (Table 11) [71].

The benzene ring containing cyclobutanones (*208*), (*210*) and (*212*) were also oxidized to their γ-lactones (*209*), (*211*) and (*213*) respectively [71]. These examples demonstrate convincingly the order of alkyl migration.

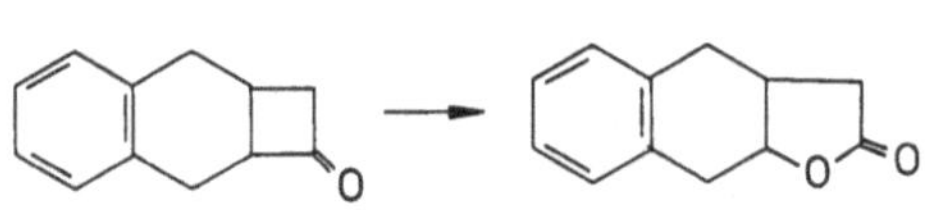

63.8 %

208 209

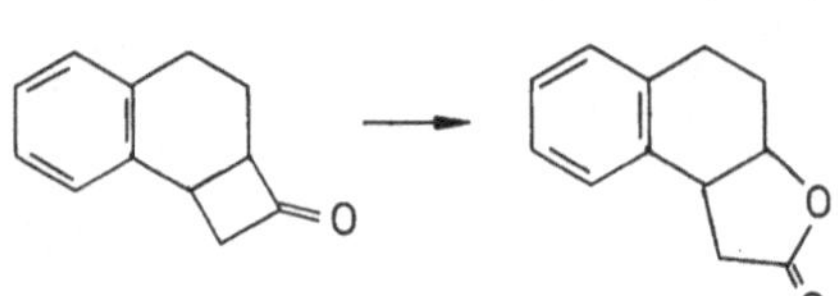

58.6 %

210 211

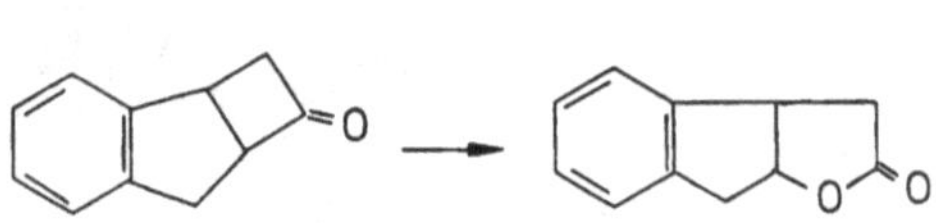

64.3 %

212 213

Table 11

	n	R	yield (%)		
			(*205*)	(*206*)	(*207*)
(*204a*)	4	H	—	—	60 from (*206a*)
(*204b*)	4	Ph	89	84	93
(*204c*)	4	Me	95	100	90
(*204d*)	4	^{i}Pr	—	46 from (*204d*)	71
(*204e*)	3	Me	90	43	78

The transformation of the 11-norPGE$_2$ (*214*) to lactome (*215*) with H_2O_2 in HOAc was achieved [72].

On the other hand, treatment of 11-nor PGE$_2$ methyl ester (*216*) with an excess of *o*-mesitylenesulfonylhydroxylamine in CH_2Cl_2 at 0 °C, followed by passage of the reaction mixture through a bed of basic alumina yielded regioselectively the γ-lactam prostaglandin (*217*) [72].

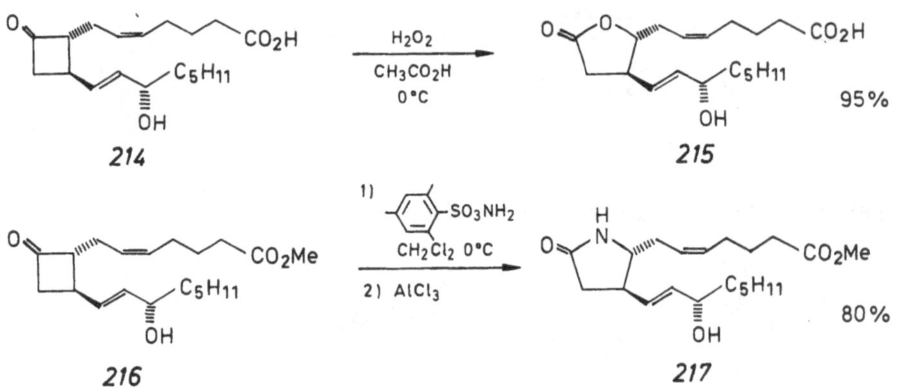

The ring expansions of cyclobutanones to cyclopentanones, γ-lactones as well as γ-lactams are perhaps the most important applications of cyclobutane derivatives. The reaction is extremely feasible due primarily to the ready availability of cyclobutanones, and particularly to the ease in the control of the insertion of methylene groups and oxygen atoms through reaction with diazomethane and peracids respectively. Surely, the strain energy of cyclobutanones plays also a very important role in the ring expansion reaction. It is therefore not astonishing to find out that the ring expansion reaction of cyclobutanones is very popular in organic synthesis.

2.4 Ring Opening by Action of Bases or Nucleophiles

Cleavage of cyclobutane rings can occur easily. As indicated in the introductory section, the strain of the four-membered ring, the substituents on the ring, the nature of the reagents as well as the conditions of reaction are all responsible for the ease of cleavage of cyclobutanes [3]. The substituents on the ring constitute one of the most important controlling factors since their presence not only facilitates the ring fission, but also help to maintain the regioselectivity of the attacking nucleophiles.

The following cleavage reactions depict the importance of these hetereoatom substituted cyclobutanes in organic synthesis.

Irradiation of the cycloalkenone (*218*) with vinylene carbonate in THF resulted in the separation of (*219*), which yielded the β,γ-unsaturated aldehyde (*220*) upon treatment with KOH in MeOH at room temperature (Table 12) [73].

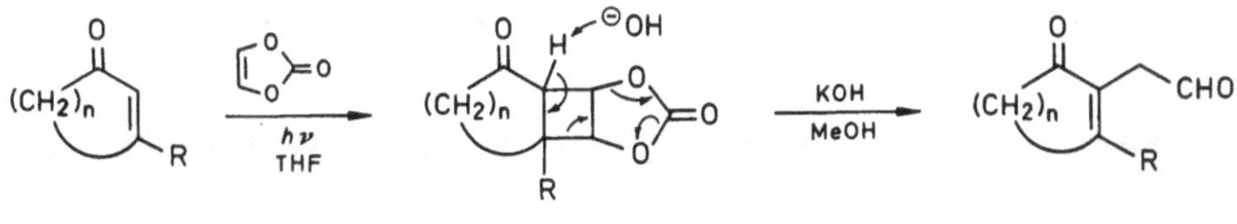

218 *219* *220*

Table 12

(*218*)		overall yield of (*220*)
n = 3	R = H	50%
n = 3	R = Me	77.5%
n = 2	R = H	40%

Isophorone (*221*) provided the aldehyde (*222*) [73].

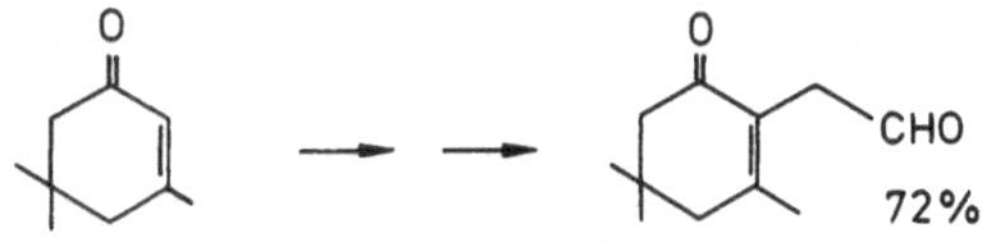

221 *222*

Treatment of RCH(SPh)COCl with triethylamine resulted in the *in situ* generation of alkyl (phenylthio) ketenes, which were trapped by olefinic compounds following thermal[2 + 2]cycloaddition to form cyclobutanes (*223*) [74].

The cleavage of the cyclobutane ring of (*223*) was achieved by reaction with KOH—KO^{t}Bu in Et_2O. The reaction proceeded with excellent stereoselectivity: the *cis*-configuration of the product (*224*) was maintained (Table 13) [74].

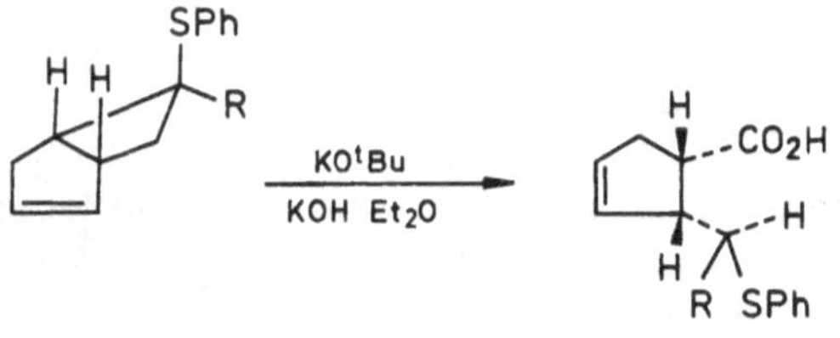

223 *224*

Table 13

	Yield of (*224*)
R = Me	93%
R = $^nC_7H_{15}$	89%
R = 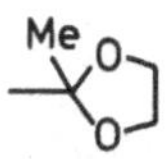	98%

The diketone (*225*) underwent cleavage reaction by a catalytic amount of NaOMe in MeOH to give the keto-ester (*226*) [75].

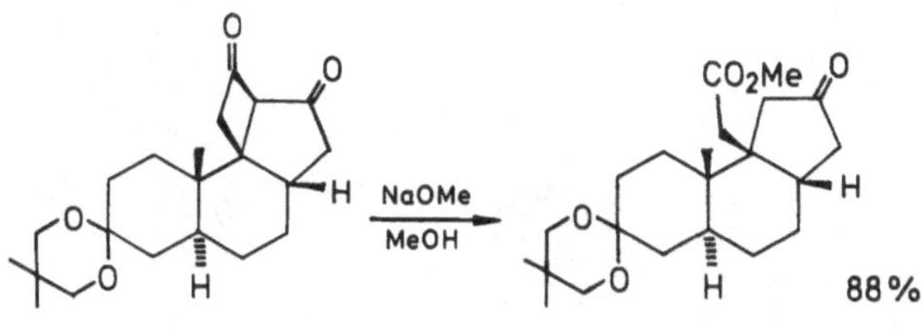

225 *226*

The key step of de Mayo's approach towards methyl marasmate made use of the rearrangement of the cyclobutane (*227*) to afford the bicyclo[2.2.1]system (*228*], which upon cleavage with $Pb(OAc)_4$ and esterification with CH_2N_2 afforded the ester (*229*) [76].

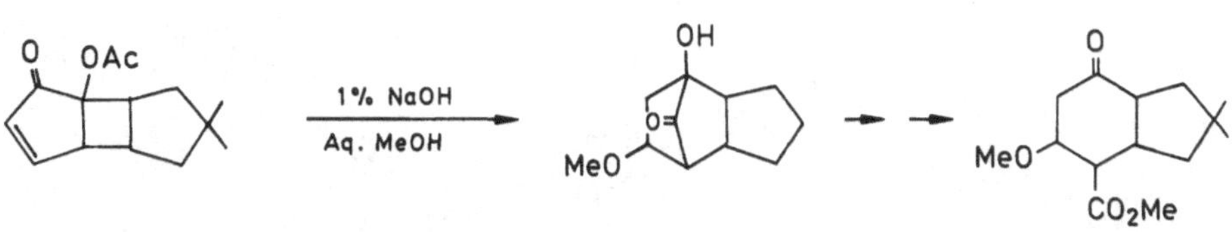

227 *228* *229*

Upon utilization of 50% aq. acetone in the presence of 2,6-lutidine as a solvolytic medium, the mesylate (*230*) underwent rearrangement to give the methyl ester of (±)-steviol (*231*) in only 3% yield [77].

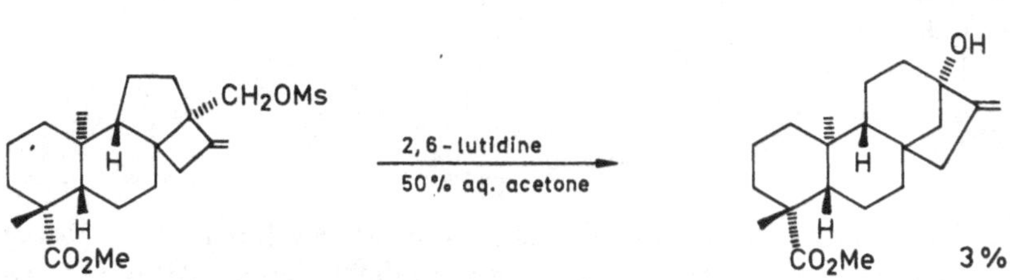

230 *231*

Treatment of cyclobutanones (*232*) with NaOMe and diphenyldisulfide in MeOH at reflux temperature led to *bis*-sulfenylation and subsequent ring cleavage to furnish compounds (*233*) (Table 14) [78].

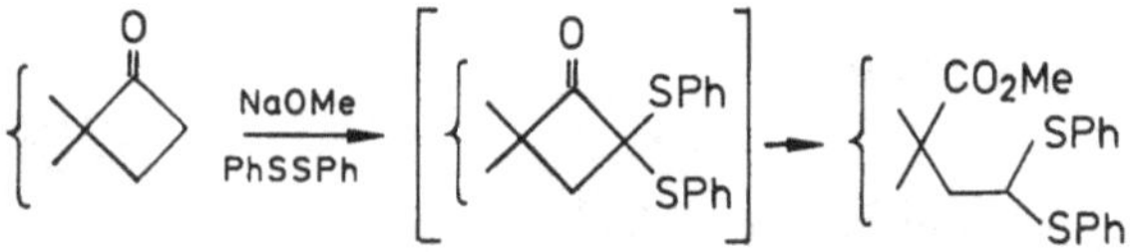

232 *233*

Table 14

(*232*)	(*233*)	Yield of (*233*)
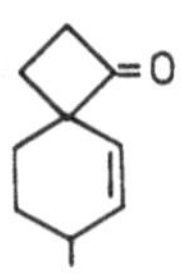	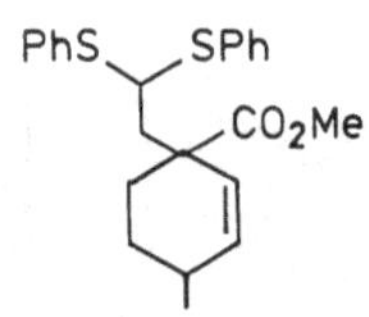	80%
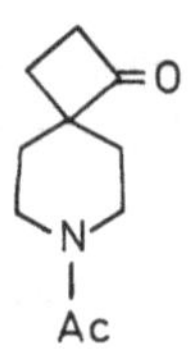	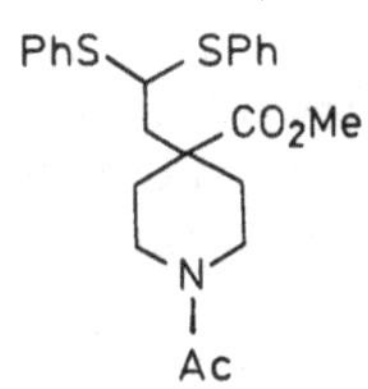	61%
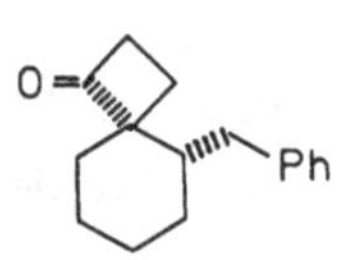	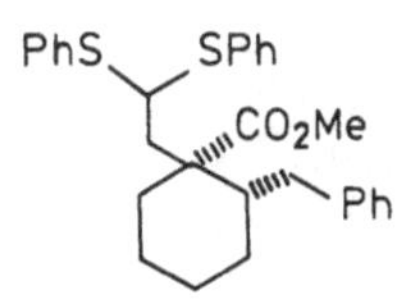	74%
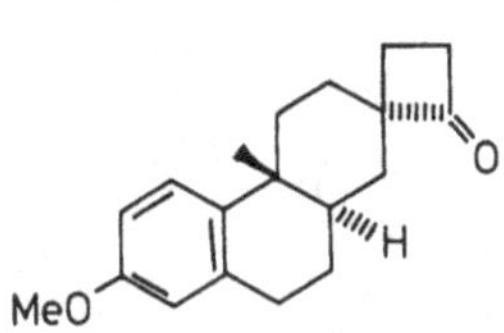	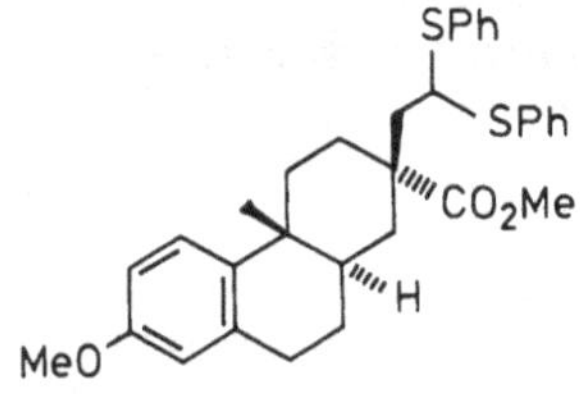	70%

Sodium methoxide catalyzed ring opening of *gem*-dichlorocyclobutanone (*234*) at -10 °C in MeOH afforded stereospecifically the *cis*-dichloroester (*235*). The *cis*-dichloroester (*235*) was reduced by $^{n}Bu_3SnH$ to the monochloroester (*236*), which could be converted to the γ-lactone (*237*) by treatment with $AgNO_3$ in aq. THF solution [79].

It is important to note that the γ-lactone (*237*) could not be obtained by the Baeyer — Villiger route [65].

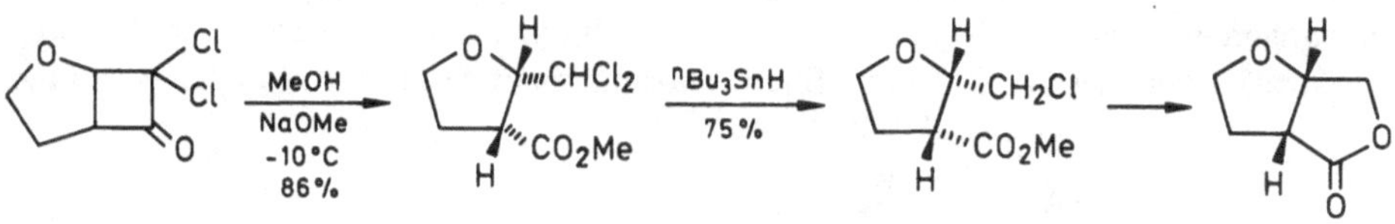

234 *235* *237* *236*

The ring opening of certain fused ring cyclobutanones led to the preparation of larger ring system. It was reported [80] that the electrophilic addition of trimethylsilyl iodide in the presence of catalyst promoted the ring opening of (*238*) (*240*) and (*242*) to afford (*239*), (*241*) and (*243*) respectivly [80].

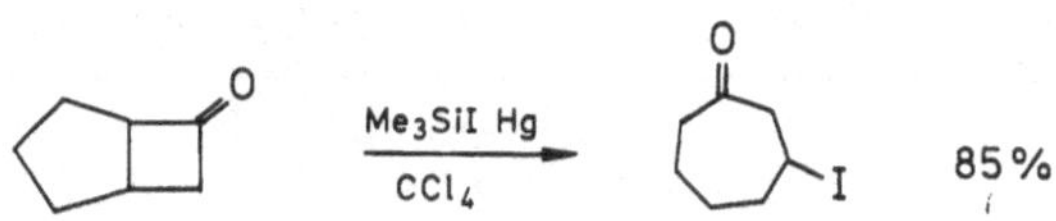

238 *239*

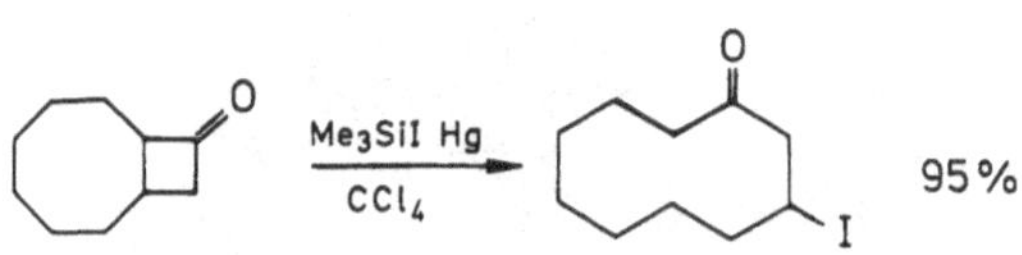

240 *241*

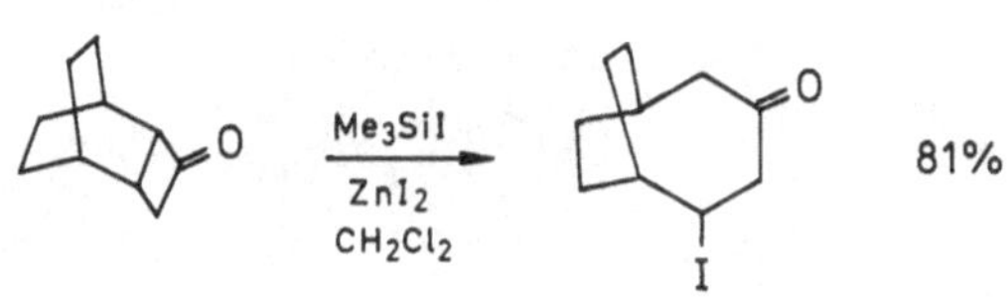

242 *243*

The increasing number of approaches that make cyclobutanones easily available from thermal [2 + 2]cycloaddition [49, 50, 51] of ketenes to olefins and polar addition of cyclopropyl ylides to carbonyls [57] greatly enhances the application of cyclobutanones as starting materials or intermediates in the construction of carbon skeletons. Because of the ring strain in cyclobutanones, the ring itself is very sensitive to the influence of nucleophiles and bases. These reagents would fission the ring regioselectivity so that skeletal reconstruction of the compounds would be attained. However, as shown by the aforementioned examples, this methodology has been applied not so widely in the synthesis of complex natural products. As a synthetic tool, this reaction has great potential because of the ready availability of starting materials and the high regioselectivity during the course of reaction.

2.5 Ring Opening and Rearrangement by Action of Acid

Acid promoted ring cleavage of cyclobutane rings inevitably involves carbenium ion intermediates. Subsequent rearrangements of these cations to more stable carbon skeletons have been utilized in organic synthesis. The most synthetically useful rearrangement reaction is known as the cyclobutylcarbinyl cation to cyclopentyl cation rearrangement (Eq. (8))[81]. Such rearrangement can proceed under very mild

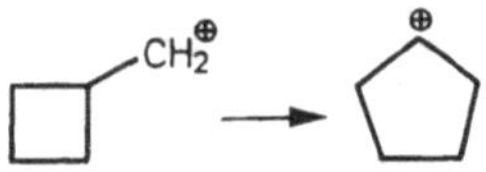

(Eq. 8)

Table 15

(244)	(245)	(246)	(247)	Rearrangement condition
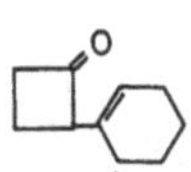			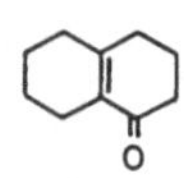 65%	$MeSO_3H/P_2O_5$ 10 : 1
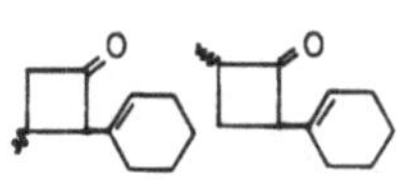			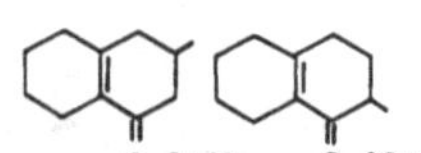 34% 32%	″
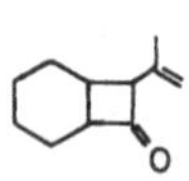			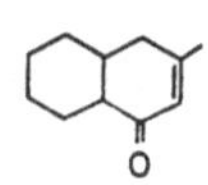 41%	″
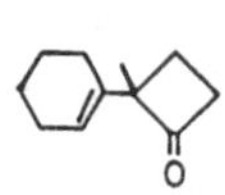	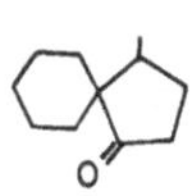 51%		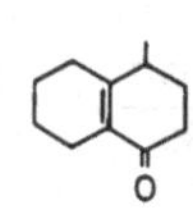 13%	″
″	53%		8%	20% $MeSO_3H$ CH_2Cl_2
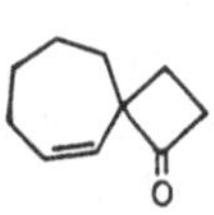		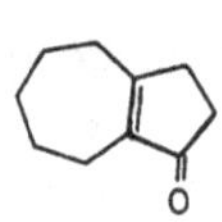 48%		$MeSO_3H/P_2O_5$ 10 : 1
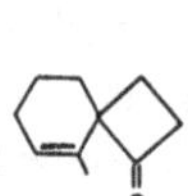	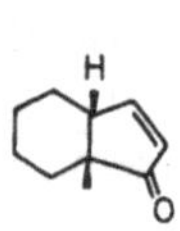 52%			$MeSO_3H$
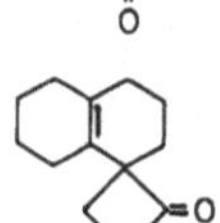	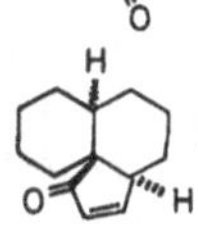 33%			$MeSO_3H$

conditions provided that the cyclobutane derivative is suitable for such molecular manipulation. The driving force for this rearrangement is again due to the ring strain of cyclobutane. The prospect of applying this rearrangement to the synthesis of complex molecules is particularly encouraging. Consequently, when the cyclobutylcarbinyl skeleton is part of a fused ring system, intriguing skeletal rearrangement can result in the isolation of novel fused ring systems. This aspect of rearrangement will be exemplified by the following examples.

Acid catalyzed the rearrangement of the 2-alkyl-2-vinylcyclobutanones (*244*) to yield either cyclopentenones (*245*), (*246*) or cyclohexenones (*247*) via 1,2-acyl migration or 1,3-acyl migration respectively (Table 15) [82].

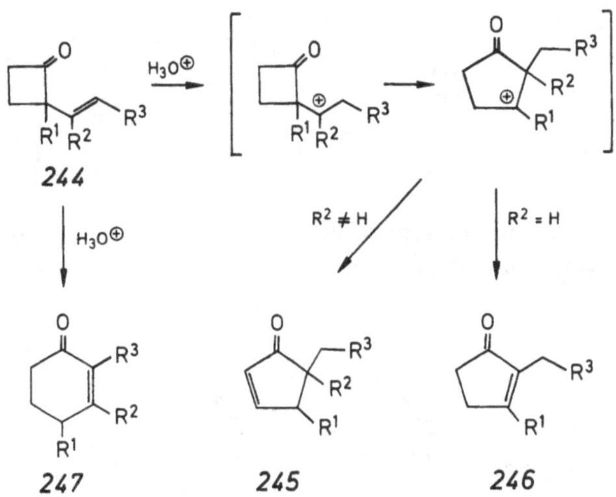

The bicyclo[3.2.0]heptene system can lead to the formation of 7-membered ring compounds. This method will be depicted by the following examples.

1,1-dimethyl-2,5-diphenyl-1-silacyclopenta-2,4-diene (*248*) underwent photo-

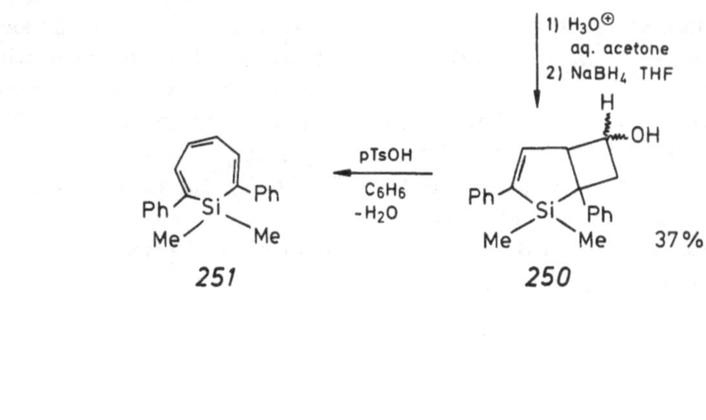

[2 + 2]cycloaddition with 1,1-dimethoxyethylene to form the adduct (*249*). Hydrolysis of (*249*) with acid and reduction of the isolable ketone with $NaBH_4$ provided the alcohol (*250*). The silacycloheptatriene (*251*) was isolated in only ca 11% yield together with other products when (*250*) was subjected to dehydration with *p*TsOH in benzene with concomitant azeotropic removal of H_2O by using a Dean-Stark trap [83].

A simple preparation of tropolone (*253*) was reported, in which the rearrangement of the 1:1 adduct (*252*) of dichloroketene and cyclopentadiene readily provide tropolone (*253*) by catalyzed with KOAc—HOAc [84].

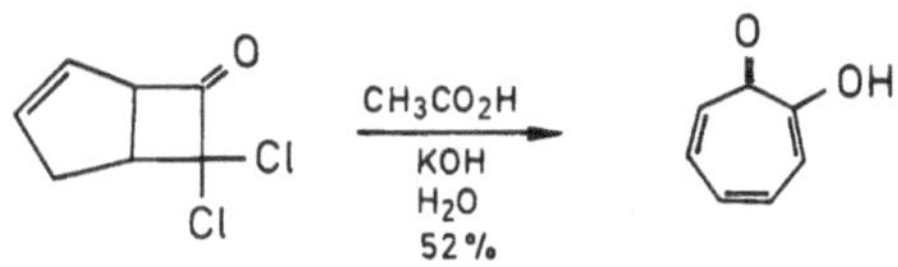

252 253

The mechanism of this rearrangement has been studied in details by Bartlett [85]. And this strategy has been applied to the synthesis of β-dolabrin (*255*) [86] as well as β-thujaplicin (*257*) [87].

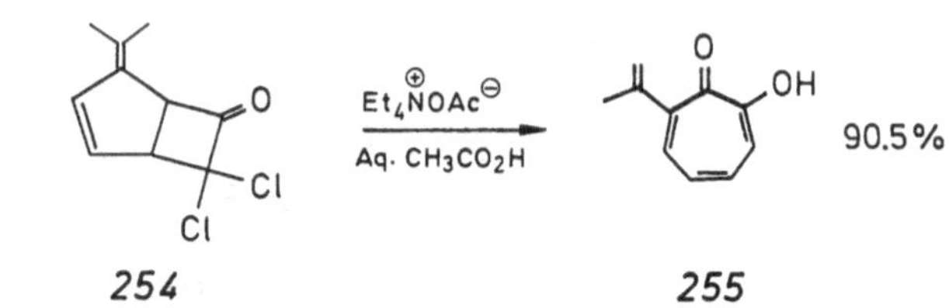

254 255

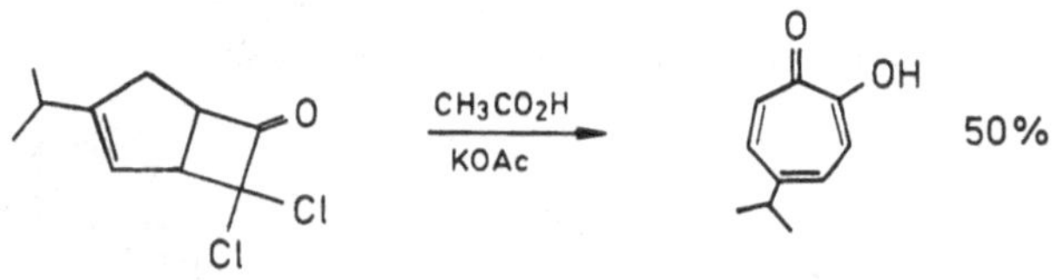

256 257

Benzotropolone has also been prepared by using the same method [88].

The cyclobutanone (*258*) reacted with acid to furnish the keto-acid (*259*). Upon esterification, ketalization and reduction, (*259*) was converted to the alcohol (*260*). Mesylation of the alcohol (*260*) and then treatment of the mesylate with NaN_3 in DMSO provided the azide (*261*). The azide (*261*) was then transformed to the urethane (*262*) by reduction and ethyl chloroformate reaction. The urethane (*262*) was deketalized by acid, nitrosated by N_2O_4—NaOAc and decomposed by NaOEt—EtOH to give the ketone (*263*) [89]. The ketone (*263*) served as a starting material for the synthesis of veatchine (*264*) [90].

The ketone (*265*) was prepared *via* photo[2 + 2]cycloaddition between the corresponding olefin and allene [91].

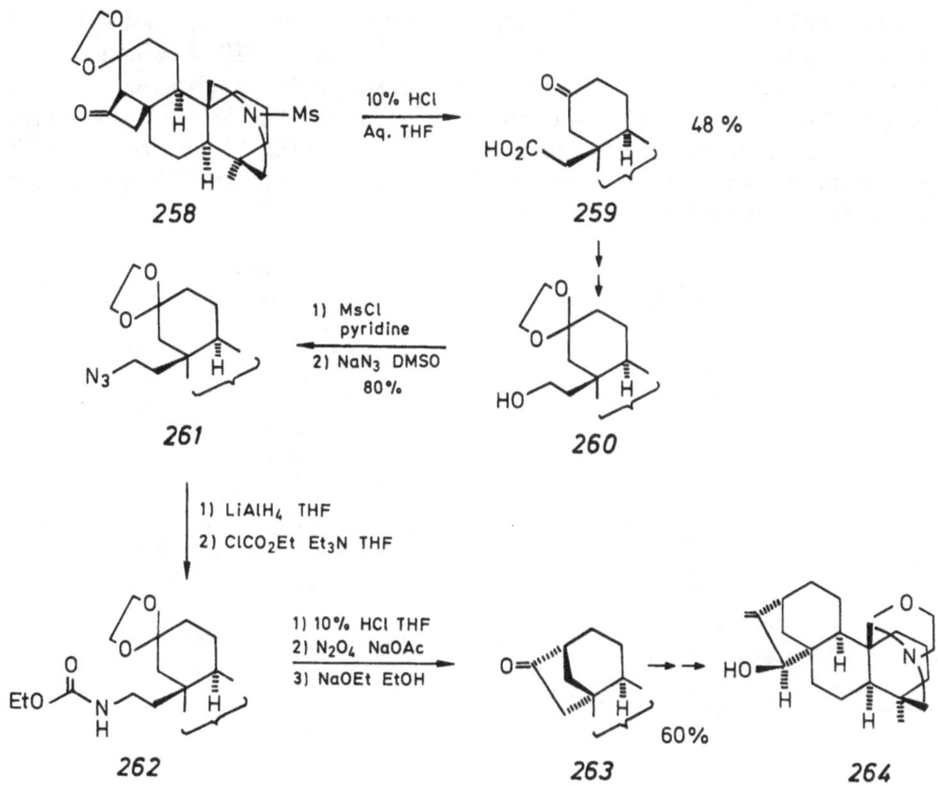

The ketone (*265*) was able to rearrange to isophyllocladenone (*267*) when heated in benzene in the presence of excess *p*TsOH, presumably via the acid catalyzed rearrangement of the intermediate (*266*) [91].

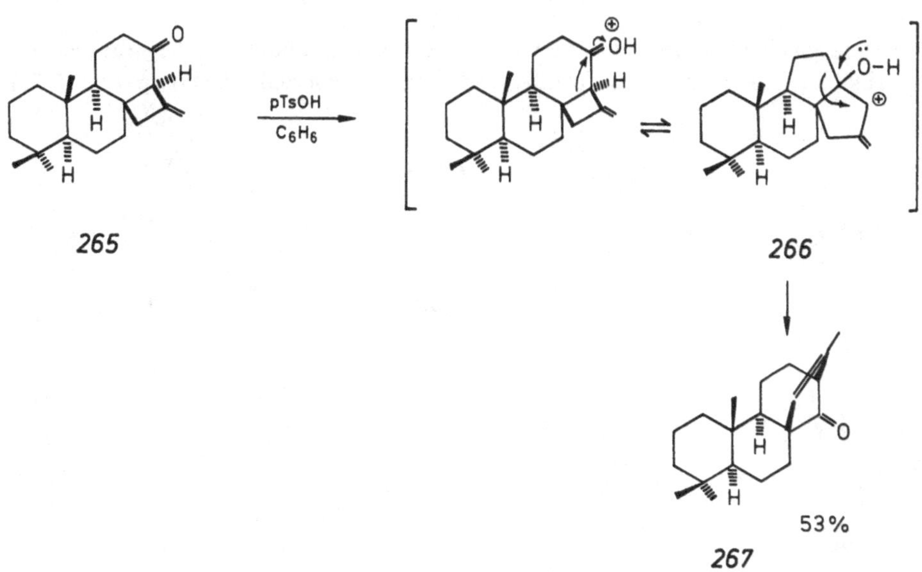

The cyclobutene (*268*) was constructed through the photo[2 + 2]cycloaddition of acetylene to anhydromevalonolactone, which was followed by DIBAL reduction [92].

A model study [92] showed that when the hemiacetal (*268*) was solvolyzed in formic acid, the bicyclic product (*269*) was isolated in good yield [92]. This method was then used to synthesize the tricyclic nucleus of verrucarol (*272*) [92]. On reaction with *p*TsOH, the acetate (*270*) underwent rapid rearrangement to (*271*), which constituted part of the structural feature of verrucarol (*272*) [92].

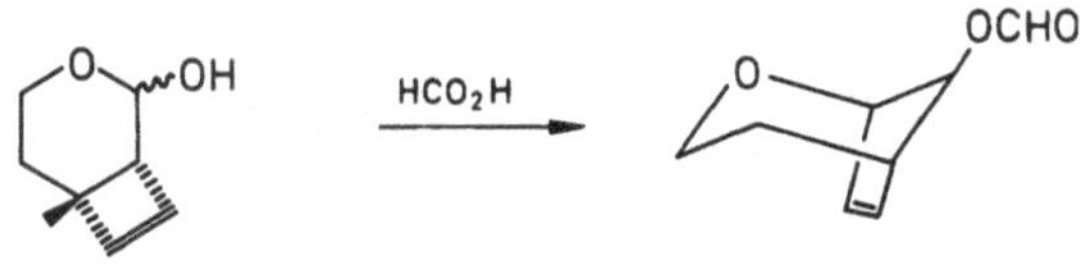

268 269

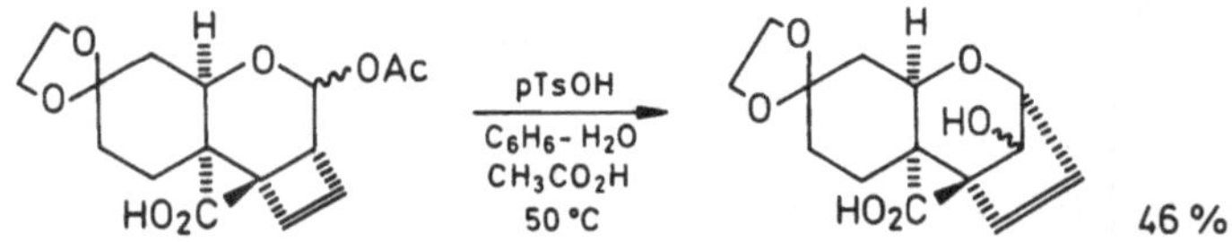

270 271

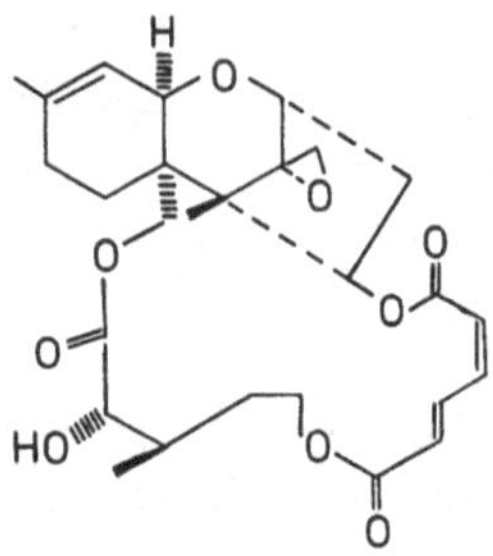

272

Another skeletal rearrangement involving the strained cyclobutane ring was recently reported [93]. The formic acid catalyzed reaction converted the compound (*273*)

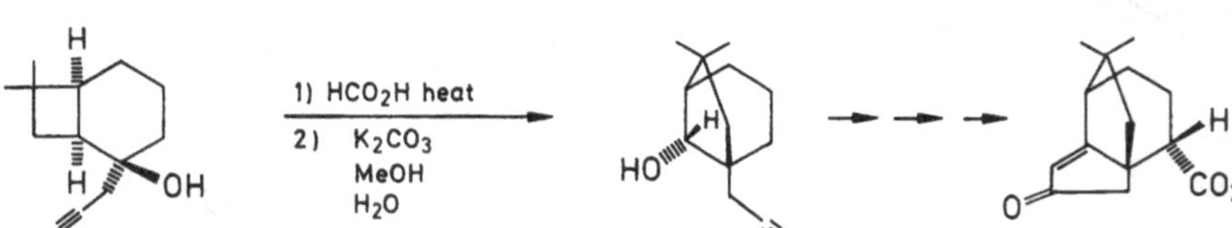

273 274 275

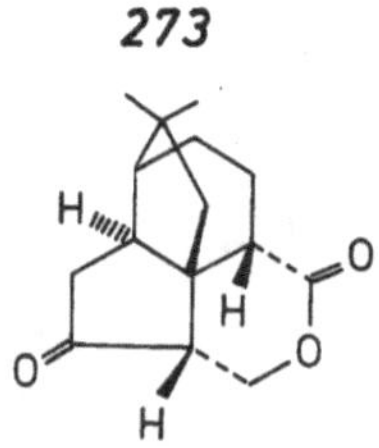

276

to the tricyclic product (*274*), which was an intermediate in the synthesis of (*275*). The acid (*275*) could be transformed to the sesquiterpene quadrone (*276*) [93]. This reaction scheme demonstrates the synthetic application of the cyclobutylcarbinyl cation-cyclopentyl cation rearrangement [81].

Cleavage of the cyclobutane ring of (*277*) by Lewis acid followed by cyclization of the allylsilane (*278*), furnished the tricyclic compound (*278*) [94].

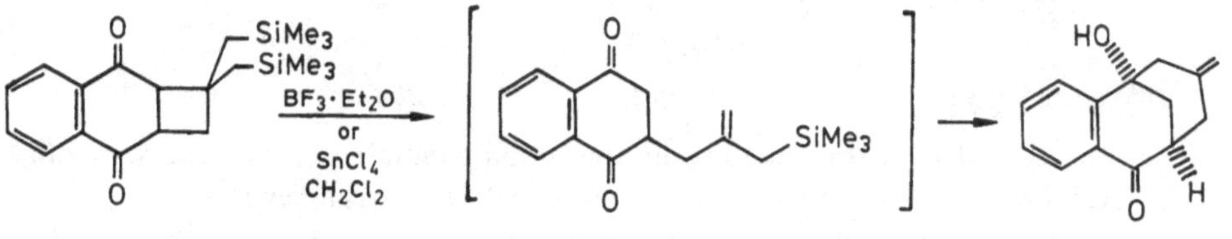

277 *278* *279*

The yields of the reaction were 28% and 32% respectively when $BF_3 \cdot Et_2O$ and $SnCl_4$ were used as catalyst [94].

When the ketone (*280*) was heated at reflux with *p*TsOH in benzene, the product (*281*) was isolated [95]. The mechanism of this intriguing rearrangement may involve 1,3-hydride shift or epoxide formation [95]. This reaction appears to be an efficient method for the synthetis of [3.3.3]propellane.

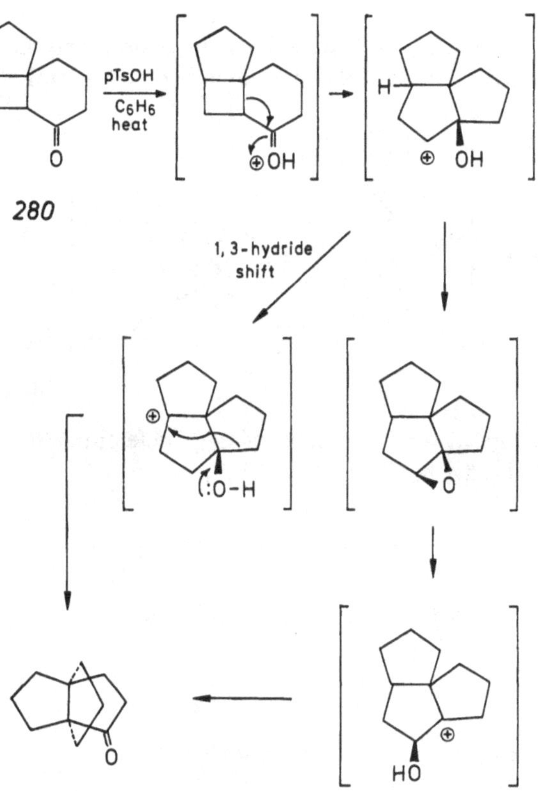

280

281

The [3.3.3]propellane skeleton could also be prepared from the cyclobutene (*282*), which provided the product (*283*). The propellane (*283*) was an intermediate compound in the synthesis of modhephene (*284*) [96].

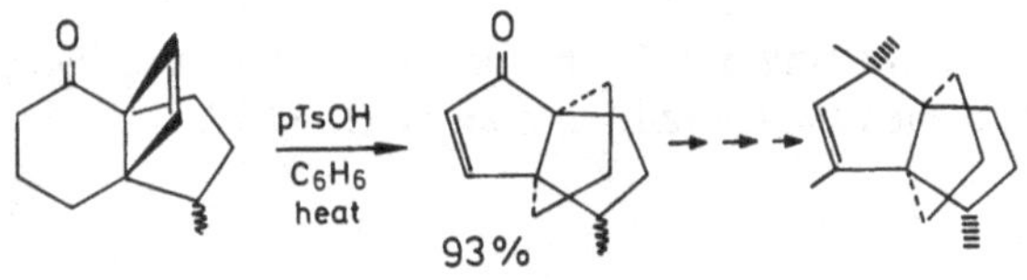

282 *283* *284*

The synthesis of (±)-isocomene is another good example showing the versatility of the cyclobutylcarbinyl cation — cyclopentyl cation rearrangement. A high yield of isocomene (*286*) was obtained when the olefin (*285*) was treated with *p*TsOH in benzene [97].

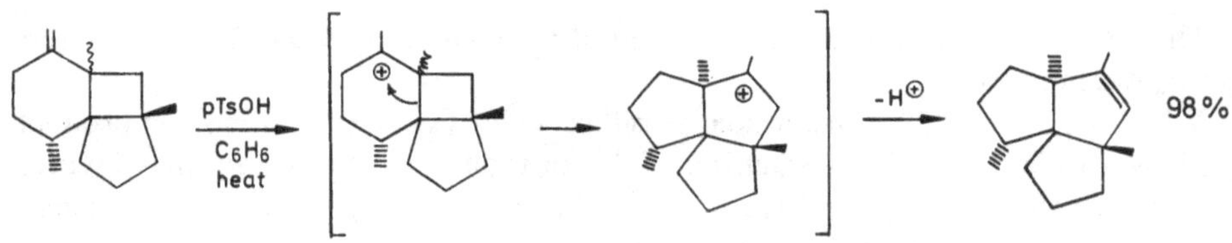

285 *286*

Treatment of the ketone (*287*) with *p*TsOH in benzene at 60 °C provided the norbornen-7-one derivative (*288*) [98], following interesting acid-catalyzed rearrangement.

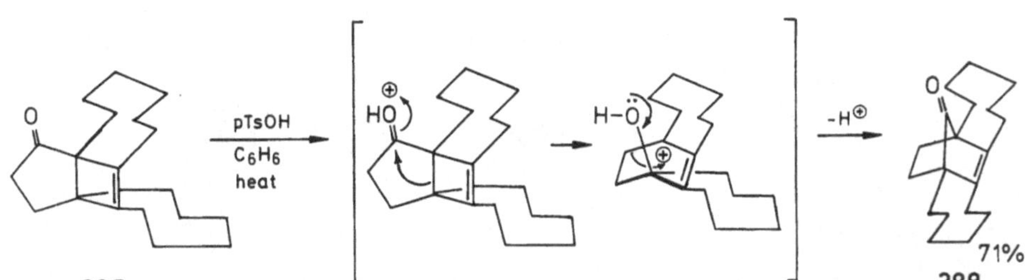

287 *288*

The key step in the formal total synthesis of (±)-brefeldin A utilized the acid catalyzed ring opening of (*289*) to (*290*) [99].

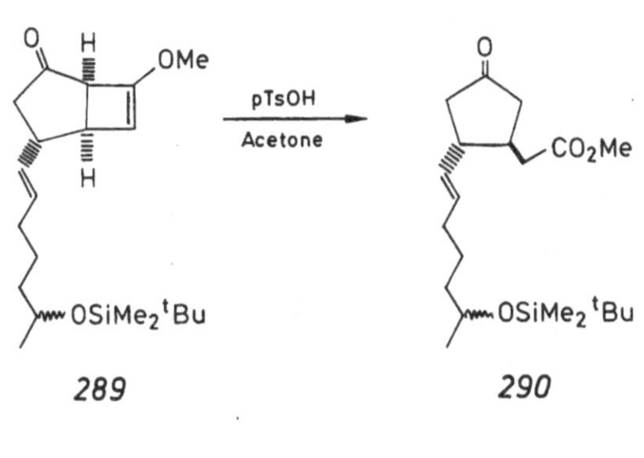

289 *290*

The diketone (*291*) could also undergo acid catalyzed ring cleavage to provide the keto-acid (*292*), which possesses the structural feature of a propellane [100].

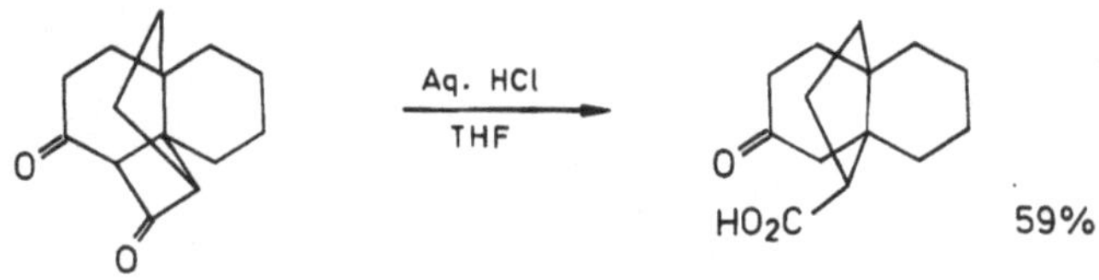

291 *292*

The norbornanes (*294*) could be efficiently prepared from the bicyclic compounds (*293*) via acid catalyzed rearrangement (Table 16) [101].

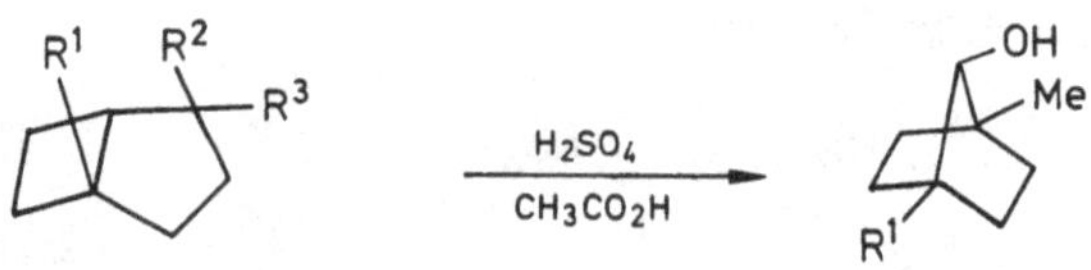

293 *294*

Table 16

R^1	R^2	R^3	yield of (*294*)
H	$-CH_2-$		98%
Me	$-CH_2-$		81%
H	OH	Me	70–93%
H	Me	OH	70–93%
Me	Me	OH	75%

Two recent reports recorded the ring opening of pinene and its derivatives [102, 103]. The acid treatment of the epoxide (*295*) provided (*296*), (*297*) as well as (*298*) [102].

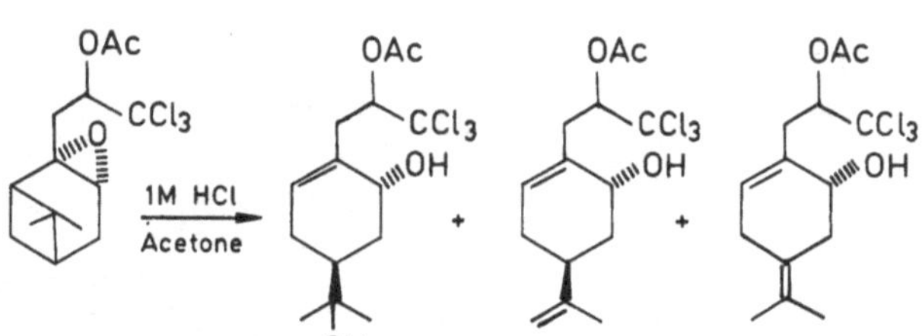

295 *296* *297* *298*

On the other hand, the (—)-α-pinene (*299*) yielded a mixture of (*300*), (*301*), (*302*),

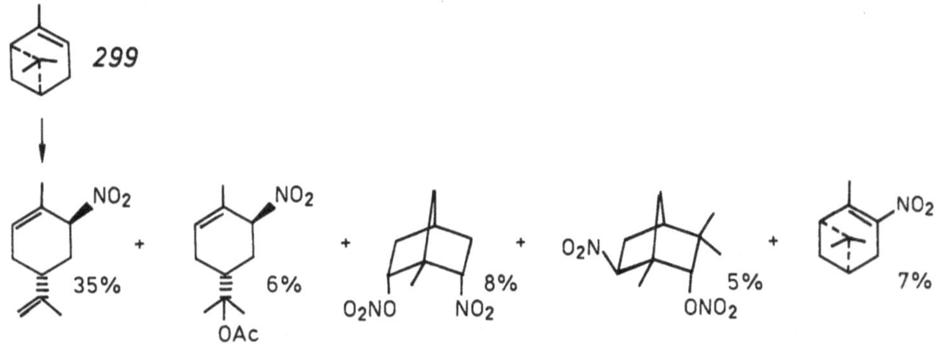

300 301 302 303 304

(*303*) and (*304*) on treatment with excess nitrous acid [103]. Similarly, (—)-β-pinene (*305*) yielded (*306*), (*307*), (*308*) and (*309*) [103].

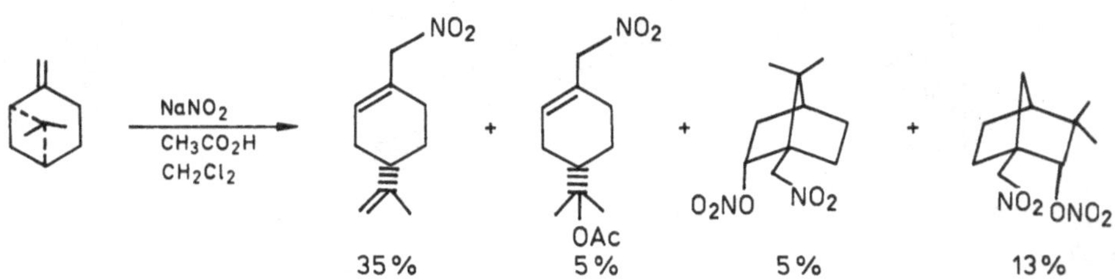

305 306 307 308 309

3 Thermal Ring Opening

3.1 In Situ Generation of 1,3-Butadiene and o-Xylylene Derivatives

The intramolecular Diels-Alder reaction has become one of the most promising methods for the construction of complex cyclohexene moieties [104]. In view of the fact that cyclobutenes rearrange thermally and conrotatary to butadienes (Eq. (9)) [81,105] and their willingness to undergo Diels-Alder reactions with dienophiles (Eq. (10))

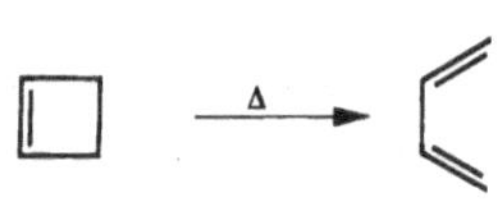

(Eq. 9)

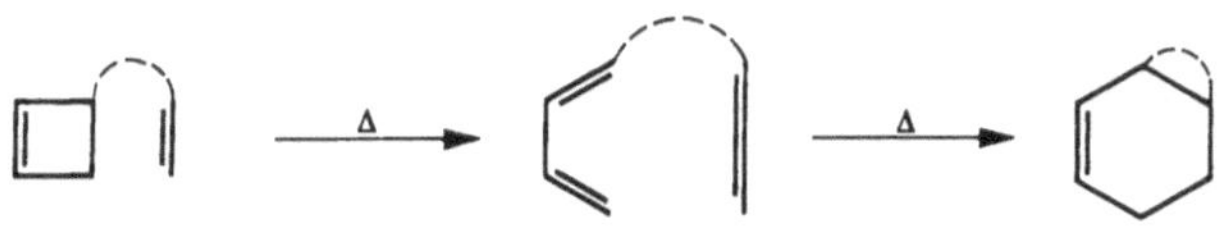

(Eq. 10)

either intermolecularly or intramolecularly, much efforts have been devoted to construct complex molecules applying this strategy [104]. The ease of the cyclobutene-butadiene rearrangement depends very much on the properties of substituents. Therefore it is not surprising that cyclobutene itself rearranges to butadiene at temperatures higher than 200 °C, yet other cyclobutenes are less demanding and yield butadienes between 100–200 °C [81]. Furthermore, the presence of a fused benzene ring on the cyclobutene framework can affect the equilibrium of the rearrangement reaction. Thus, upon heating, benzocyclobutenes isomerize to *o*-xylylenes [81] (Eq. (11)), which can be trapped by dienophiles intermolecularly as well as intramolecularly (Eq. (12)) [81, 104, 106].

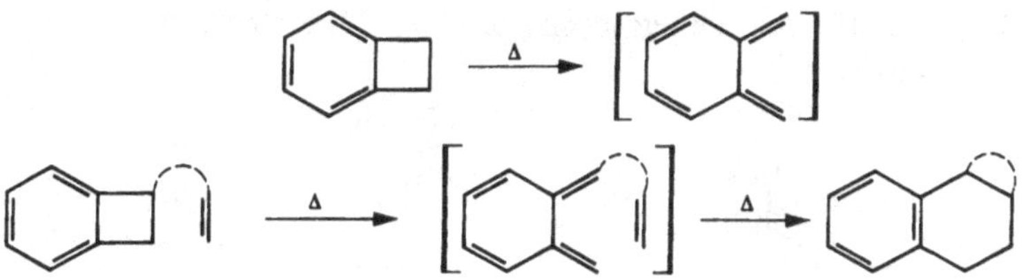

(Eq. 11)

(Eq. 12)

The forward reaction is extremely easy bevause of aromaticity recovery. These aspects of the intramolecular Diels-Alder reaction are generally very useful and able to provide polycyclic fused six-membered ring compounds which are otherwise difficult to realize. The controlling factors, geometry and mechanism of intramolecular Diels-Alder reactions have been comprehensively reviewed elsewhere [104, 106], and it is not our intention to discuss these in details. However, the synthetic utility of the reaction is demonstrated by the following examples [107].

The cyclobutene (*310*) opened at 100 °C to facilitate the intramolecular Diels-Alder reaction, from which the *cis*-fused bicyclic compound (*311*) was isolated [108].

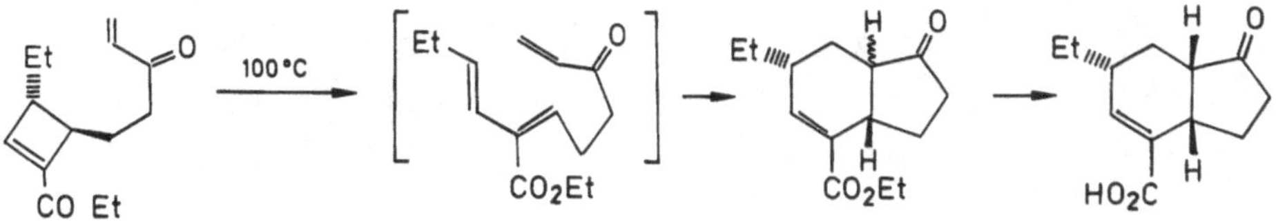

310 *311*

Although substituent effects on the conrotatary ring opening of cyclobutenes to butadienes have not been studied systematically, empirical findings reveal that the temperatures required for the rearrangement range from 100 °C to 200 °C. The parent compound, cyclobutene, which required temperature higher than 200 °C for its rearrangement, is perhaps the only exception. Consequently, substituents should in practice more or less lower the reaction temperature of the cyclobutene — butadiene rearrangement provided that they do not exert too much steric effect on the conrotatary process. However, it has been recorded that some substituents can lower the reaction temperature to as low as —30 °C. Thus, the extraordinary accelerating power of the arylsulfoxy and sulfonyl carbanion substituents on cyclobutene rings (*312*) was demonstrated by their ready opening to the butadienes (*313*) [109].

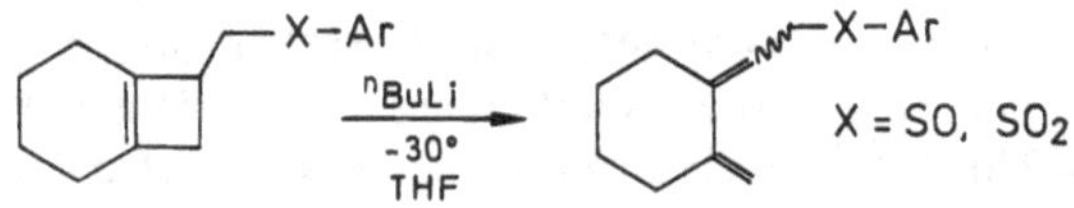

312 *313*

On the other hand, upon heating, the ketone (*314*) gave the intermediate (*315*) at temperatures between 150–170 °C. The diene (*315*) would react with 1,4-naphthoquinone (*316*) and afforded the tetracyclic ketone (*317*) after treatment with oxygen and sodium methoxide [110].

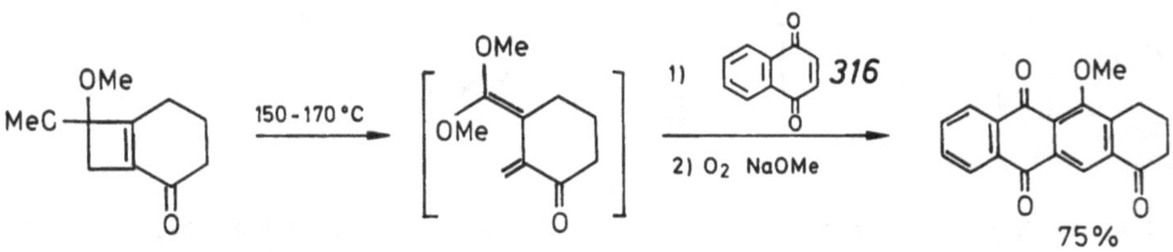

314 *315* *317*

The thermal reaction of the compound (*318*) may be chosen as an example to show the usefulness of the benzocyclobutene — *o*-xylylene rearrangement in alkaloid synthesis [111]. On heating (*318*) at 120 °C, an intermediate product (*319*) was formed, which cyclized to give the compound (*320*). This method formulated the key step in the total synthesis of the alkaloid chelidonine (*321*) [111].

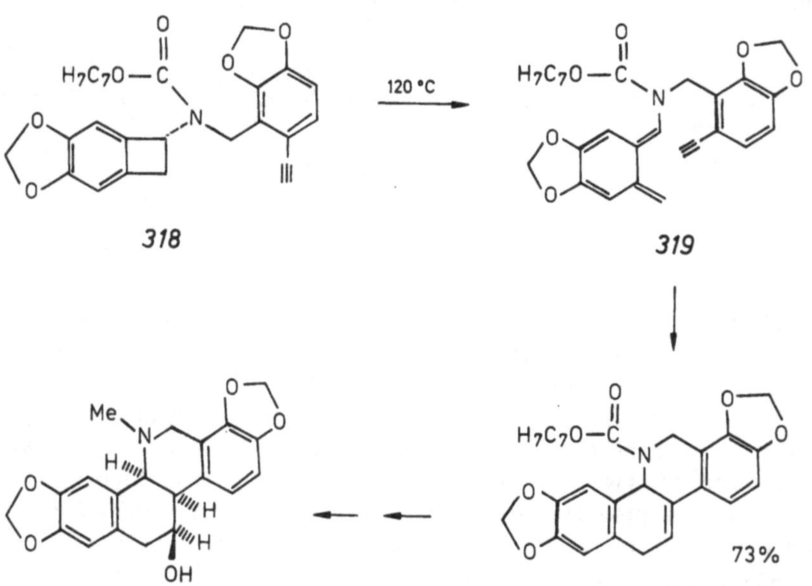

318 *319* *321* *320*

Although benzocyclobutenes have been shown to be versatile starting materials in the synthesis of polycyclic compounds by virtue of their tendency to open to give *o*-xylylenes, the most fatal drawback of this method is the relative difficulty of constructing substituted benzocyclobutenes. It has been suggested by Vollhardt that oligomerization of alkynes to benzene derivatives may be an ideal solution to this problem. Thus, cyclopentadienylcobalt dicarbonyl promoted oligomerization of acetylenes (*322*) and (*323*) led to the intermediate benzocyclobutene (*324*), which opened on heating to (*325*). Intramolecular Diels-Alder cycloaddition of (*325*) afforded the cyclic ether (*326*) [112].

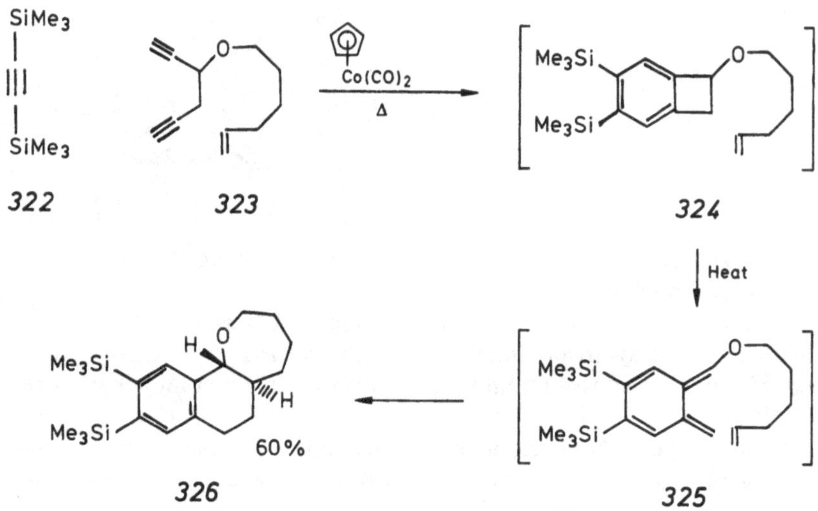

A toluene solution of the benzocyclobutene (*327*) was thermolyzed by heating in a sealed-tube at 180–200 °C for 48 hours to yield the lactone (*328*), via an intriguing epimerization pathway [113], in which the initially formed kinetically controlled *trans* product was isomerized to the charged intermediate and then recyclized to furnish the thermodynamic *cis*-isomer (*328*) [113].

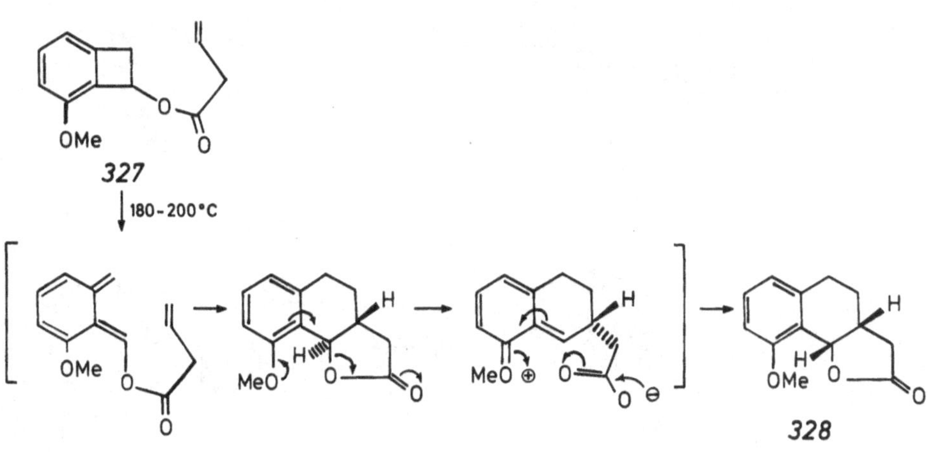

Steroid molecules can be constructed by applying the benzocyclobutene-*o*-xylylene method. When the compound (*329*) was heated in *o*-dichlorobenzene at 180 °C for 14 hours, the compound (*330*) was isolated, which could be converted by a number of steps to the steroid acetate (*331*) [114].

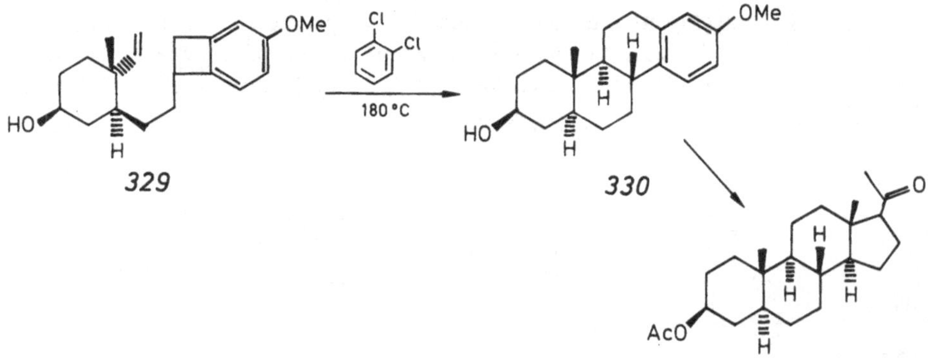

329 330 331

Total syntheses of steroids continue to be one of the most severe challenges to the skill and stamina of synthetic chemists.

The steroid estrone (*332*) contains a benzene ring and thus provides the best structural feature for the synthetic application of the benzocyclobutene-*o*-xylylene method. Indeed, many successful applications of this methodology have been recorded [115–118].

From these examples, one should recognize that the building of the complex steroid skeleton of estrone has been made rather simple when various *o*-xylylene generation and trapping methods are applied [115–118].

The approaches of Grieco [115] and Kametani [116] were similar. The benzocyclobutene (*333*) was thermally opened and then subsequent intramolecular Diels-Alder reaction afforded (*335*), which was converted eventually to (±)estrone (*332*) [115,116] in fair yields.

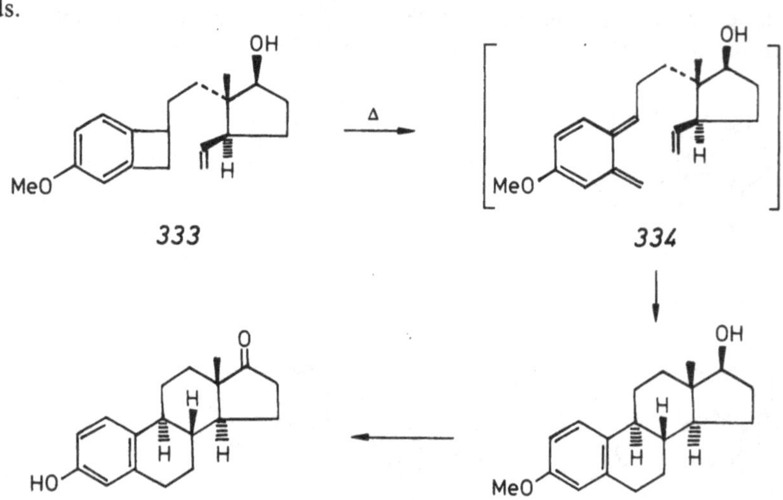

333 334 332 335

Nicolaou's approach [105] was somewhat different. The sulfone (*336*), upon heating, lost SO_2 and formed the benzocyclobutene (*337*) as an intermediate. Ring opening of (*337*) would lead to a steroid framework, which was then converted to (±)-Estrone (*332*) [117].

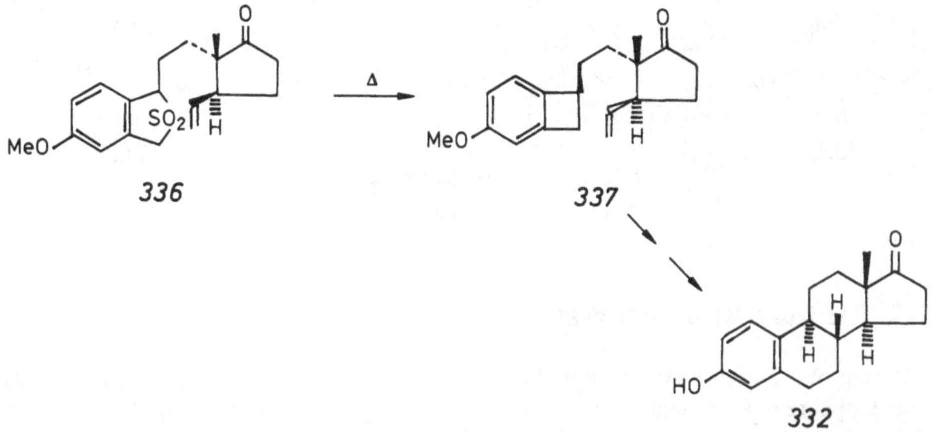

Vollhardt [118] employed the organocobalt-promoted reaction between *bis*(trimethylsilyl)acetylene (*322*) and the diyne (*338*) to construct the steroid estrone [118].

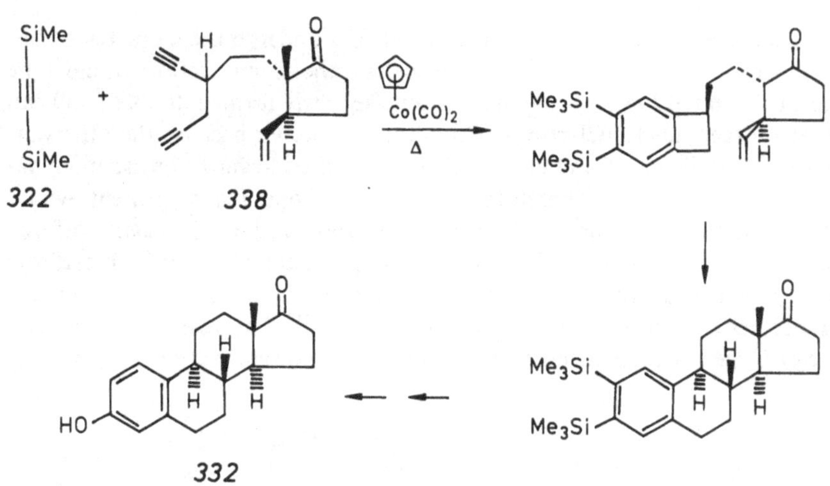

As indicated by the aforementioned examples [115–118], the most useful feature of the intramolecular trapping of *o*-xylylenes generated from benzocyclobutenes is the control of stereochemistry so that high purity of (±)-estrone could be obtained. Subsequent molecular manipulation of synthetic estrone may give rise to alternative routes to other medicinally important steroid drugs.

The thermal ring opening of the cyclobutene in the heterocycle (*339*) led also to cyclohexene derivatives (*340*) through intermolecular Diels-Alder reaction [119].

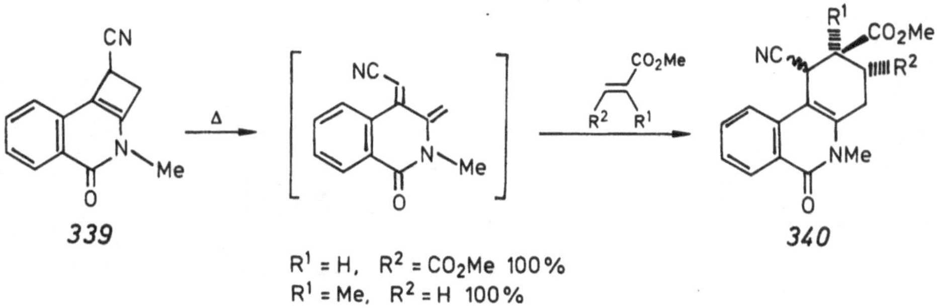

3.2 The Cope Rearrangement

Heating of 1,5-dienes results in a [3,3]sigmatropic rearrangement which is known as the Cope rearrangement (Eq. (13)) [120]. The starting 1,5-diene and the product 1,5-diene will be identical when X=H and they are different when X is anything but

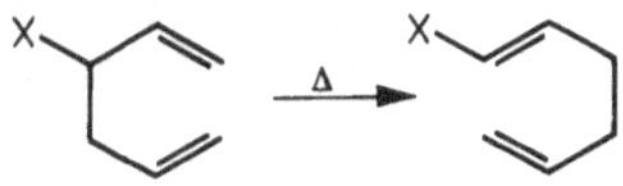

(Eq. 13)

hydrogen. It has been found that almost all 1,5-dienes undergo the Cope rearrangement at temperatures around 300 °C. Nevertheless, the isomerization would take place more easily and at lower temperature when the newly formed double bond can conjugate with other functional group. This situation occurs because the otherwise reversible reaction is disturbed and the equilibrium is shifted towards the thermodynamically more stable isomer. Therefore, although a Cope rearrangement would produce an equilibrium mixture of both reactant and product, it usually affords the thermodynamically more stable isomer in larger quantity. Furthermore, the rearrangement is irreversible for 3-hydroxy-1,5-dienes, because of the product's ability to tautomerize to carbonyl compounds (Eq. (14)) [121]. This reaction is generally called the oxy-Cope rearrangement, and it requires a far lower reaction temperature.

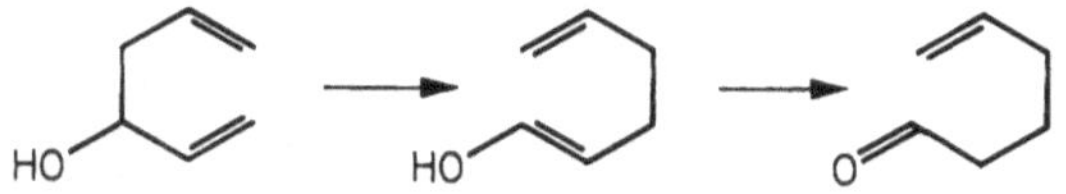

(Eq. 14)

The Cope and oxy-Cope rearrangement are very useful in organic synthesis, particularly when the 1,5-diene system is incorporated in a ring, then intringuing cyclic compounds may result. Also, the Cope and oxy-Cope rearrangements are greatly facilitated for a *cis*-1,2-divinylcyclobutane (Eq. 15)), resulting in the formation of an eight membered ring compound. The following reactions would demonstrate the versatility of the Cope and oxy-Cope rearrangement.

(Eq. 15)

Photocycloaddition of allene to cyclohexenone (*341*) gave the β,γ-enone (*342*), which reacted with vinyl magnesium bromide to produce the tertiary alcohol (*343*) in 79% yield. When the compound (*343*) was treated with KH and 18-crown-6 in THF at room temperature for two hours and quenched with aq. NH_4Cl, the cyclobutene (*344*) was obtained. The thermal ring opening of the cyclobutene (*344*) proceeded in toluene in a sealed-tube at 180 °C for twelve hours to give a readily separable 5:1 mixture of the *cis*-olefin (*345*), and the *trans*-olefin (*346*) respectively in 95% yield. Moreover, (*345*) could be converted to a mixture of (*346*) and (*345*) in the ratio of 10:1 by irradiation. The compounds (*345*) and (*346*) possess the skeleton of the germacranes (*347*), (*348*) and (*349*) [122].

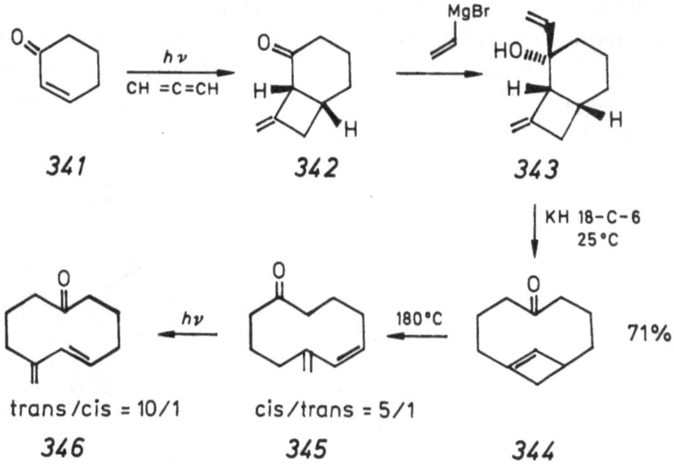

341 342 343

346 345 344

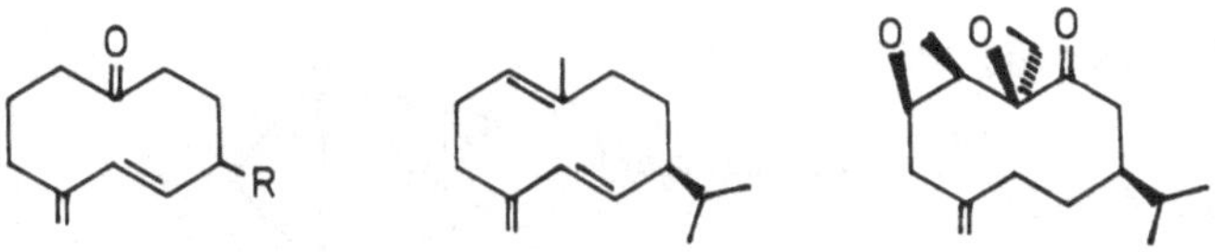

347 348 349

Rearrangement of compound (*350*) under mild conditions (NaH, refluxing THF, 30 min) gave the alcohol (*351*) as the major product, together with only a trace amount of the compound (*352*). On the other hand, rearrangement of compound (*350*) under more vigorous conditions (NaH, refluxing THF, 1.75 hrs), furnished the hydrindenone (*352*) in 10% yield [123])

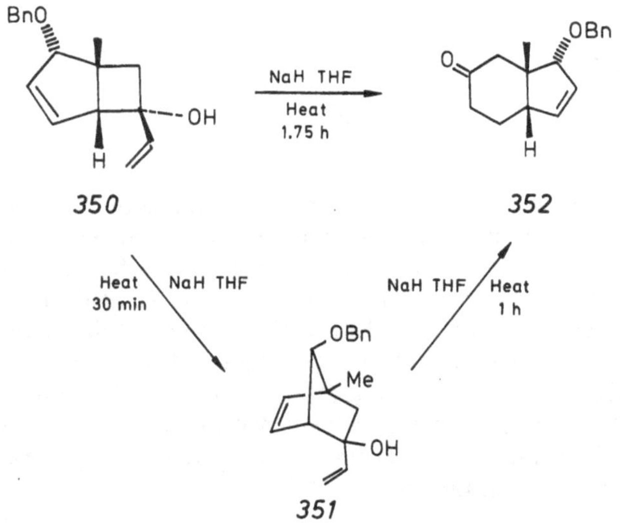

The bicyclic ketone (*353*) was treated with cyclopentenyllithium (*354*) at −78 °C to form the intermediate (*355*), which underwent a rapid Cope rearrangement to the intermediate (*356*). By treatment of (*356*) with methyl iodide, compound (*357*) was obtained. The ketone (*357*) would serve as an intermediate in the synthesis of ophiobolin A (*358*) [124]).

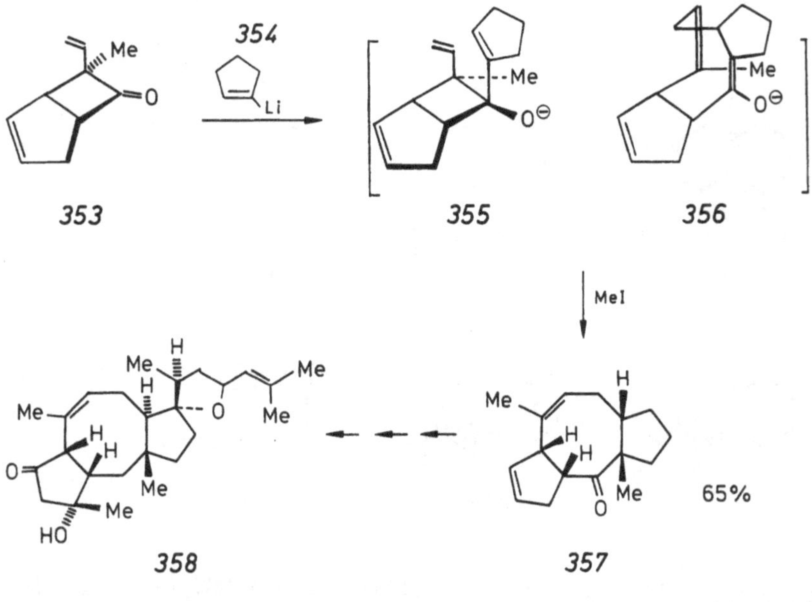

The cyclobutenone (*359*) or the α,β-unsaturated acid chloride (*360*) could be converted to the vinyl-ketene (*361*), which reacted readily with 1,3-butadiene (*362*)

to furnish 2,3-divinylcyclobutanone (*363*). At elevated temperature, (*363*) underwent Cope rearrangement to afford the eight-membered ring product (*364*). This procedure was applied to synthesize a number of eight-membered carbocycles (*367*) (Table 17) [125].

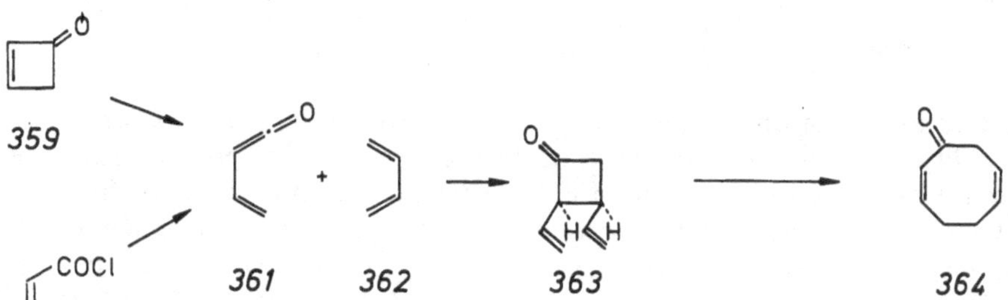

359 360 361 362 363 364

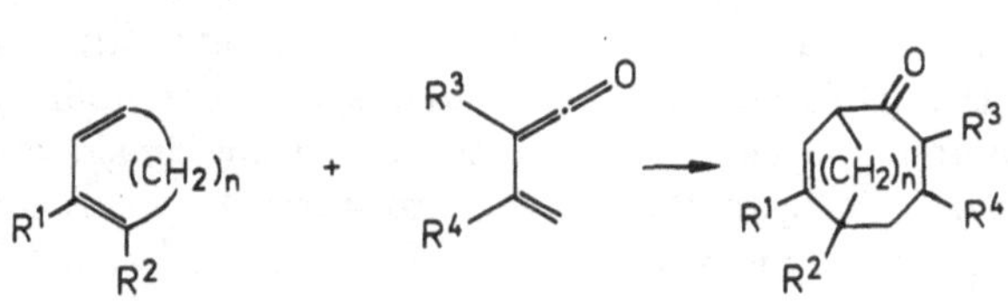

365 366 367

Table 17

n	R^1	R^2	R^3	R^4	yield of (*367*)
0	H	Me	Me	H	30%
0	$-(CH_2)_3-$		Me	H	19%
1	H	H	H	H	18%
1	H	H	Me	H	40%
2	H	H	H	Me	49%
2	H	H	H	^nBu	33%
2	H	H	Me	Me	91%
2	H	H	Me	H	31%
2	H	H	Cl	H	44%

3.3 Olefin Metathesis

When alkenes are allowed to react with certain catalysts (mostly tungsten and molybdenum complexes), they are converted to other alkenes in a reaction in which the substituents on the alkenes formally interchange. This interconversion is called metathesis [126]. For some time its mechanism was believed to involve a cyclobutane intermediate (Eq. (16)). Although this has since been proven wrong and found that the catalytic metathesis rather proceeds via metal carbene complexes and metallocyclobutanes as discrete intermediates, reactions of olefins forming cyclobutanes, which are subsequently cleaved again with formal redistribution of the substituents as in (Eq. (16)) are frequently referred to as olefin metathesis.

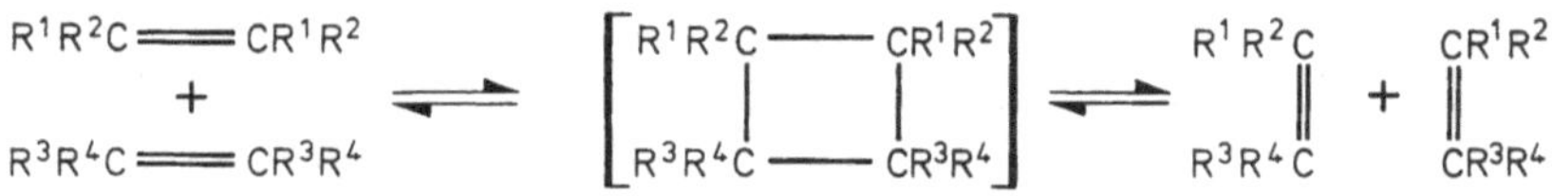

(Eq. 16)

If the cycloaddition and cycloreversion steps occurred under the same conditions, an equilibrium would establish and a mixture of reactant and product olefins be obtained, which is a severe limitation to its synthetic use. In many cases, however, the two steps can very well be separated, with the cycloreversion under totally different conditions often showing pronounced regioselectivity, e.g. for thermodynamic reasons (product vs. reactant stability), and this type of olefin metathesis has been successfully applied to organic synthesis. In fact, this aspect of the synthetic application of four-membered ring compounds has recently aroused considerable attention, as it leads the way to their transformation into other useful intermediates. For example aza[18]annulene (*371*) could be synthesized utilizing a sequence of [2 + 2] cycloaddition and cycloreversion. (*369*), one of the dimers obtained from cyclooctatetraene upon heating to 100 °C, was transformed by carbethoxycarbene addition to two tetracyclic carboxylates, which subsequently lead to the isomeric azides (*368*) and (*370*). Upon direct photolysis of these, (*371*) was obtained in 25 and 28 % yield, respectively [127]. Aza[14]annulene could be synthesized in a similar fashion [128].

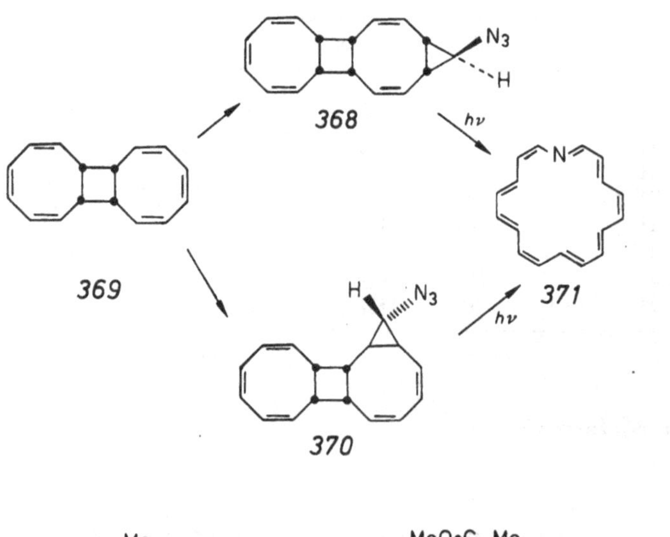

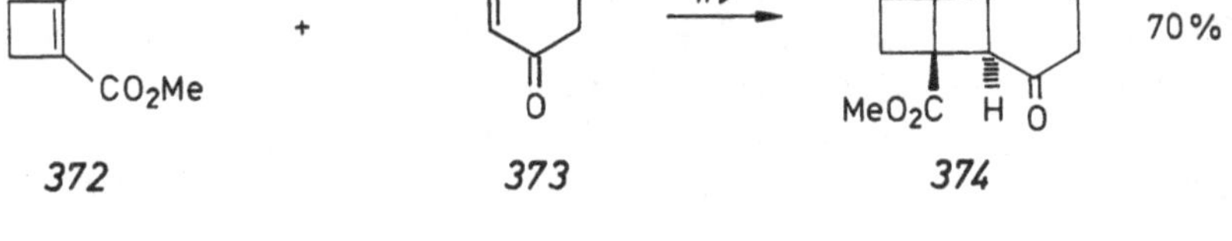

372 *373* *374*

Irradiation of a mixture of dimethyl cyclobutene-1,2-dicarboxylate (*372*) and 3-methyl-2-cyclohexenone (*373*) gave the adduct (*374*), which was then pyrolyzed to give the diene (*375*). Hydrogenation afforded the keto diester (*376*). On the other hand, when the compound (*374*) was reduced by $NaBH_4$, γ-lactone (*377*) was

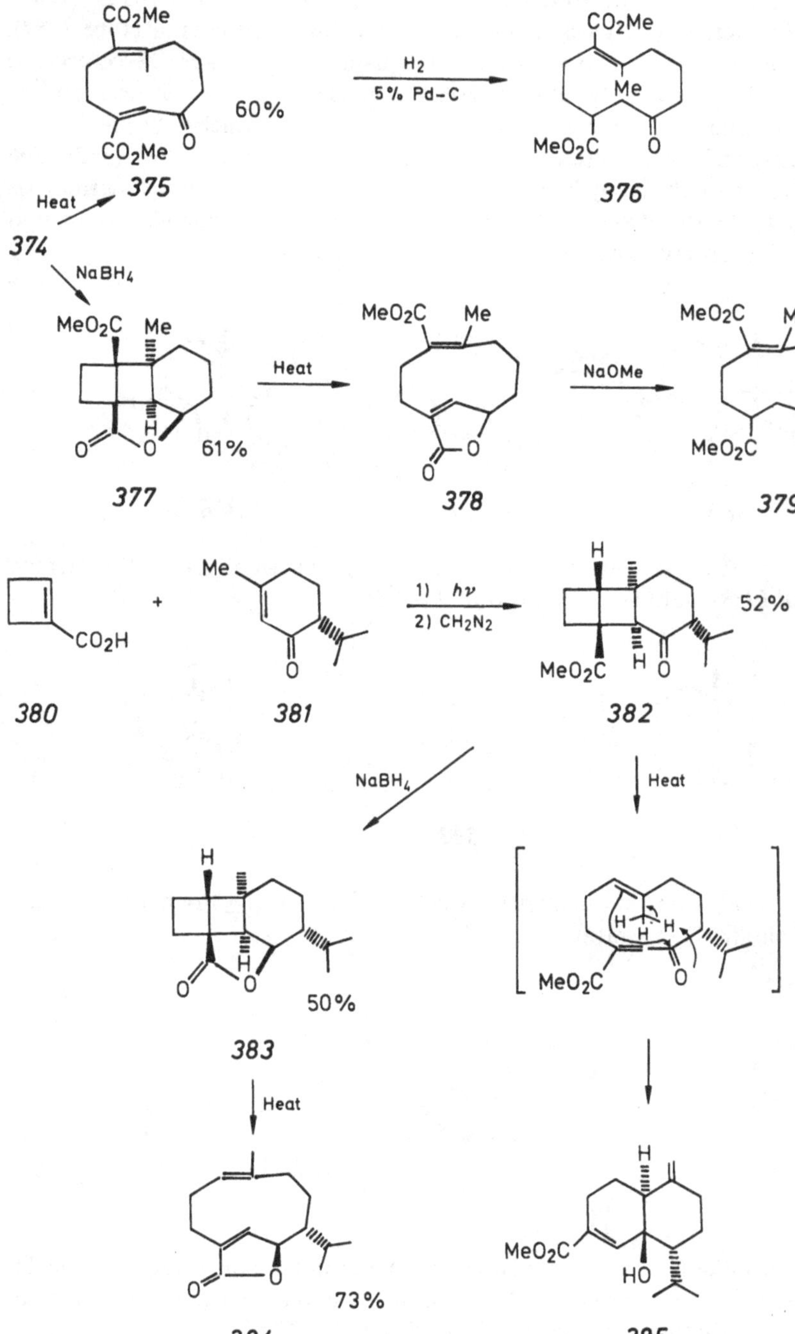

obtained. On thermolysis at 139 °C, the compound (*377*) was converted to the diene (*378*), which could also be transformed to the keto diester (*379*) [129].

The photocycloaddition of an excess of cyclobutenecarboxylic acid (*380*) and (—)-piperitone (*381*), followed by esterification with diazomethane, gave the adduct (*382*). Reduction of (*382*) afforded the lactone (*383*). Upon thermolysis in decane at 174 °C, the lactone (*383*) was converted to the dihydroisoaristolactone (*384*). Thermolysis of (*382*) in refluxing decane gave quantitative yield of the compound (*385*), presumably via a [2 + 2]cycloreversion and transannular ene type reaction [130].

An almost identical synthetic work was carried out [131], in which the photolysis of piperitone (*381*) and methyl cyclobutenecarboxylate yielded the adduct (*382*). The keto ester (*382*) was reduced by $NaBH_4$ to the lactone (*383*), which upon thermolysis at 180 °C, gave the dihydroisoaristolactone (*384*) and the elemane-like compound (*386*) [131]. Similar observations were also made by Williams [132].

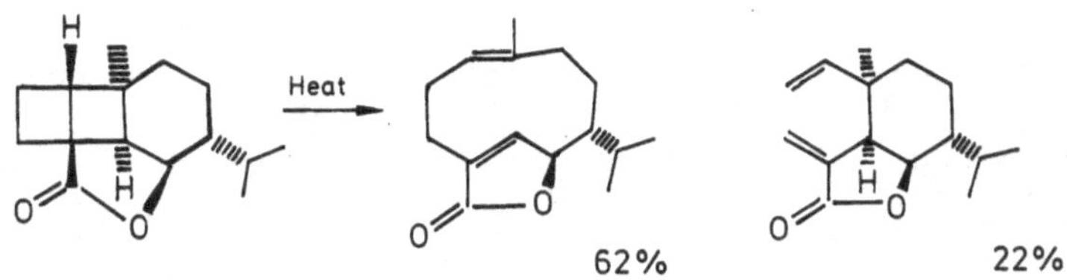

383 384 386

Wender also reported similar procedures [133–136]. Thermolysis of the lactone (*387*) afforded (*388*) and (*389*) in a ratio of 2:1 [133].

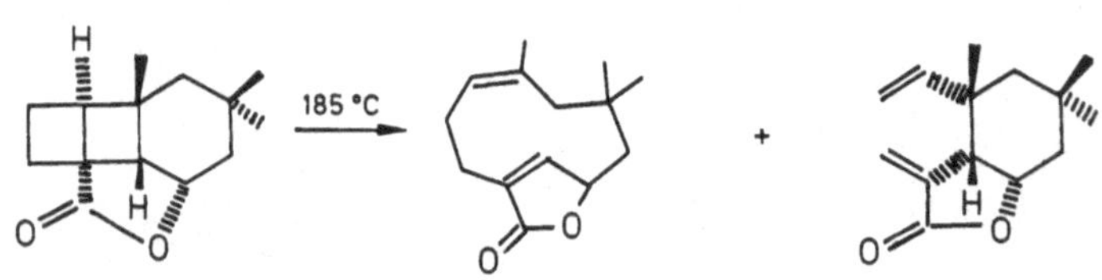

387 388 389

Pyrolysis of (*390*) gave, in quantitative yield, a mixture of (±)-isabelin (*391*) and (±)-pyroisabelin (*392*) in a ratio of 1:2 [134].

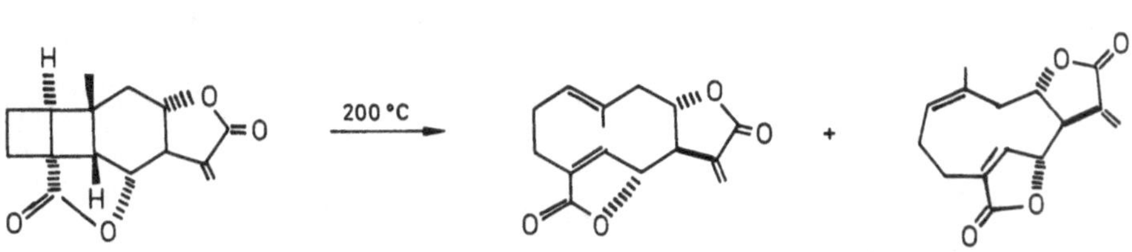

390 391 392

The olefin metathesis — transannular ene sequence usually provides *trans*-decalin structure. When (*393*) was heated at 206 °C in toluene, the ester (*396*) was isolated, presumably going through the intermediates (*394*) and (*395*) [135].

Similarly, the compound (*397a*) and (*397b*) could be converted accordingly to (*398a*) and (*398b*) in 78% and 100% yield respectively [136].

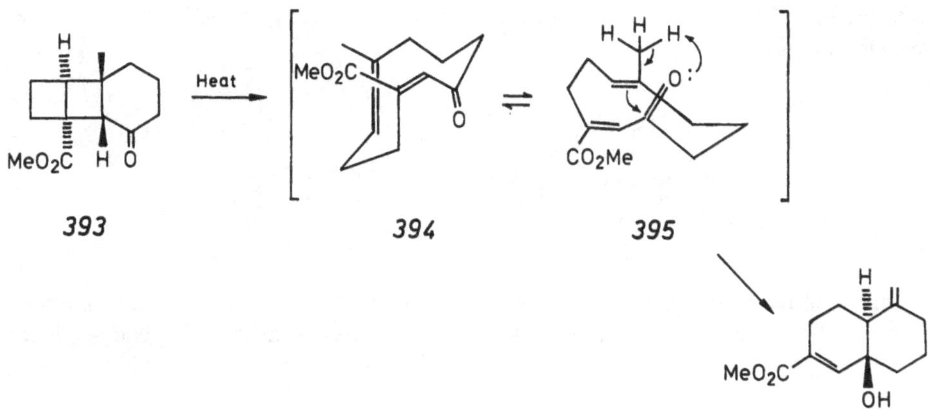

393 *394* *395* *396*

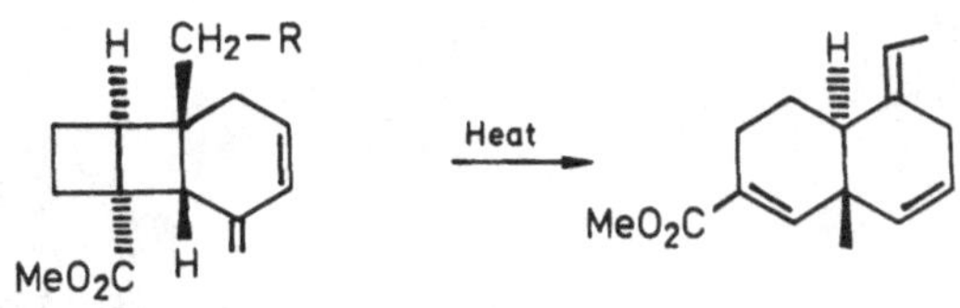

397a R = H
397b R = Me

398a R = H
398b R = Me

When no ester group is present, the thermolysis reaction may give rise to complex mixture. This is aptly explicated by the following reactions [137–139].

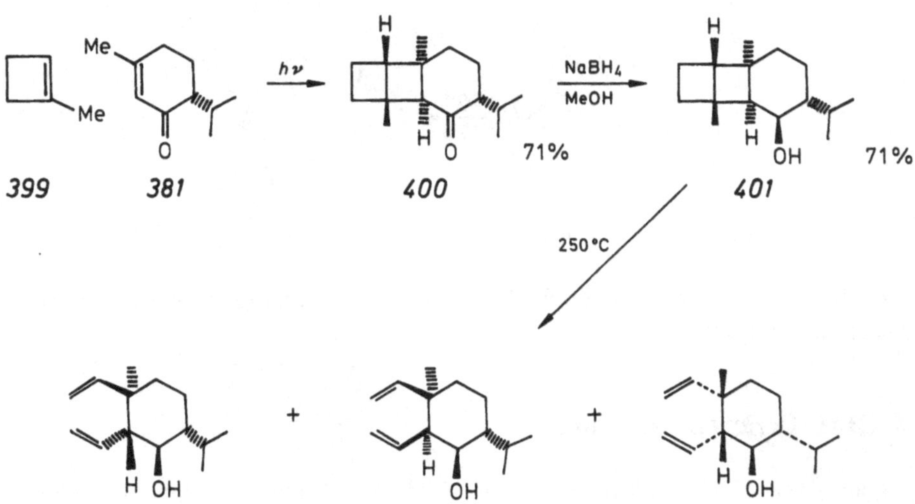

399 *381* *400* *401*

402 *403* *404*

Methyl cyclobutene (*399*) and (—)piperitone (*381*) were irradiated at -78 °C to afford the adduct (*400*). The ketone (*400*) was reduced by $NaBH_4$ to the alcohol (*401*). Thermolysis of (*401*) yielded the elemane alcohols (*402*, (*403*) and (*404*) (relative ratio 21:26:54, 48% yield), together with the aldehyde (*405*) (43%) and the germacrene (*406*) (6%) [137,139].

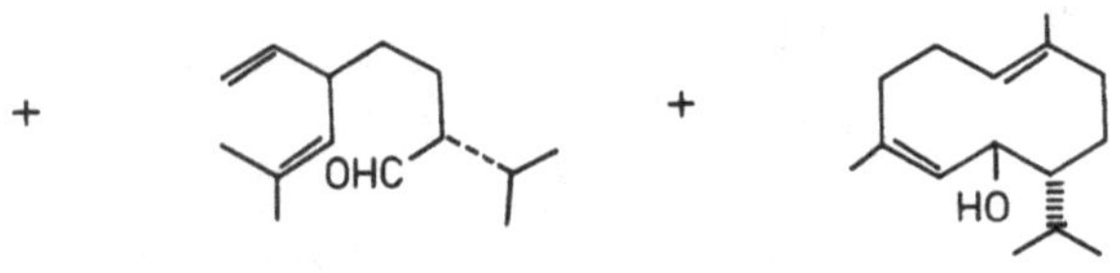

405 406

On the other hand, thermolysis of (*400*) at 250 °C in xylene afforded the cadinane dienol (*407*) in 40% yield, via also the olefin metathesis-transannular ene sequence [138,139].

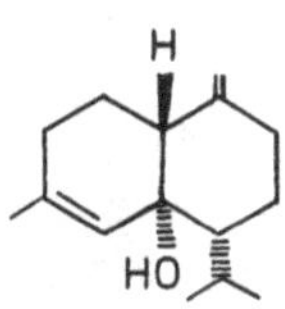

407

The key steps in Mehta's synthesis of (±)-$\Delta^{9(12)}$-capnellene (*408*), also constitute a formal olefin metathesis reaction.

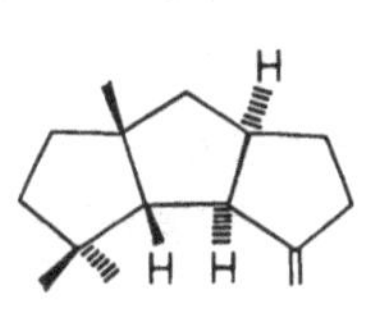

408

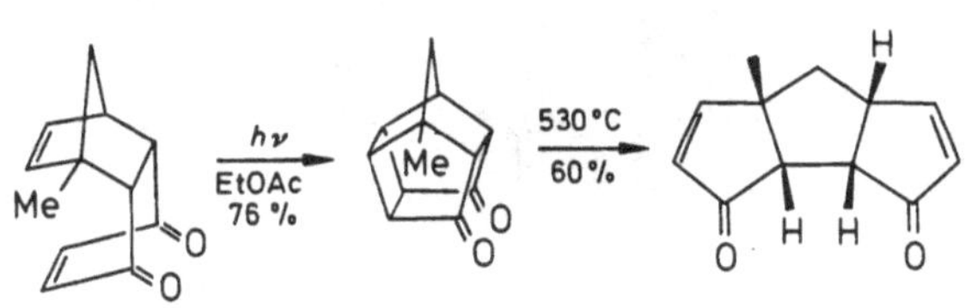

409 410 411

The diketone (*409*) cyclized photochemically to (*410*) and (*410*) was reopened to (*411*) on thermolysis [140].

3.4 Other Pericyclic Reactions

Benzyne (*412*) reacted with homoazulene (*413*) to provide benzo[*b*]homoheptalene (*415*), presumably via the rearrangement of the [2 + 2]adduct (*414*) [141].

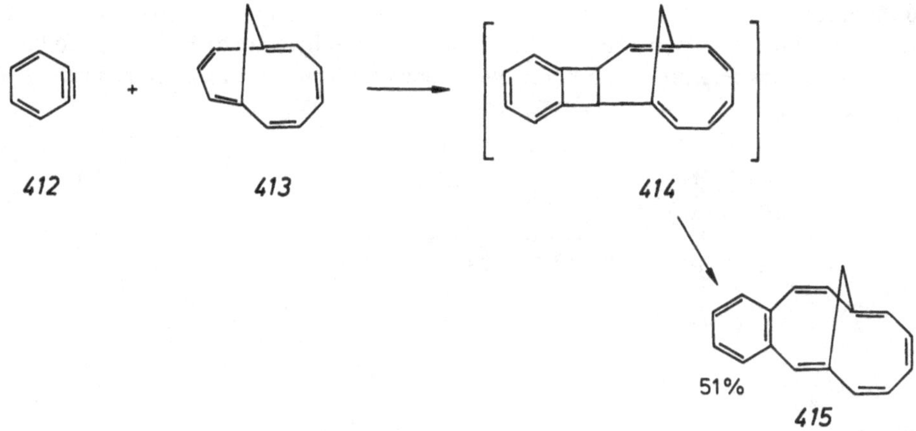

412 413 414 415

The alkoxide generated by KH in THF was believed to be effective in accelerating the vinylcyclobutane ring expansion in the synthesis of 6-membered ring compounds [142]. As an example, the cyclobutanol (*416*) reacted with KH and rearranged to (*417*), which upon subsequent oxidation, provided (*418*) [142] in 64.5% yield. The α,β-unsaturated ketone (*418*) was converted to (—)-β-selinene (*419*) [142]. Similarly, furancyclohexanol (*421*) could be obtained from the cyclobutanol (*420*) [142].

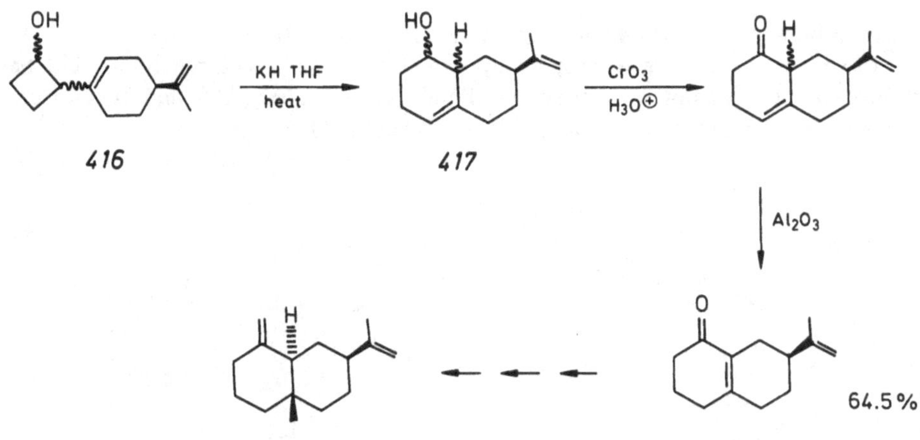

416 417 419 418

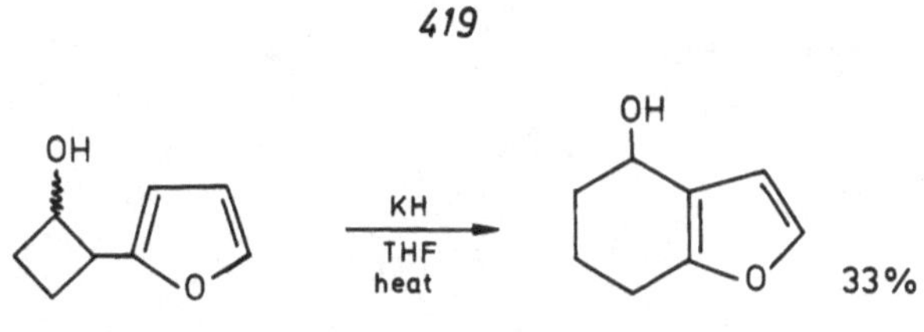

420 421

An acetone solution of 4-methoxycoumarin (*422*) in the presence of a large excess of isobutene was irradiated at >300 nm to give the cycloadduct (*423*). The compound (*423*) was allowed to react with $BF_3 \cdot Et_2O$ in benezene at room temperature to lead to the formation of 1,1-dimethyl-1,2-dihydrocyclobuta[*c*]coumarin (*424*). Finally, the compound (*424*) was converted to 4-isopropenyl-3-methylcoumarin (*425*) by refluxing in *o*-dichlorobenzene [143]. A complete review on this type of reaction has recently appeared [144].

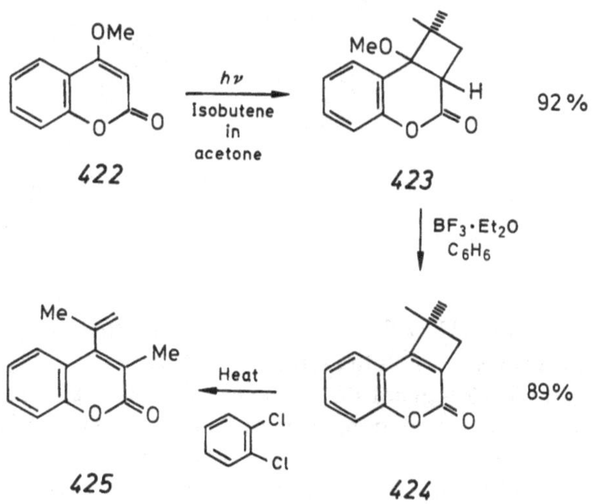

A cyclobutane ring-opening of the photoadduct (*426*) by the reaction of $BF_3 \cdot Et_2O$ in refluxing benzene gave isopropenylcyclohexenone (*427*). The reaction could be applied to more complex compounds. Thus, (*428*), (*430*), (*432*) and (*434*) were converted accordingly to (*429*), (*431*), (*433*) as well as (*435*) [145].

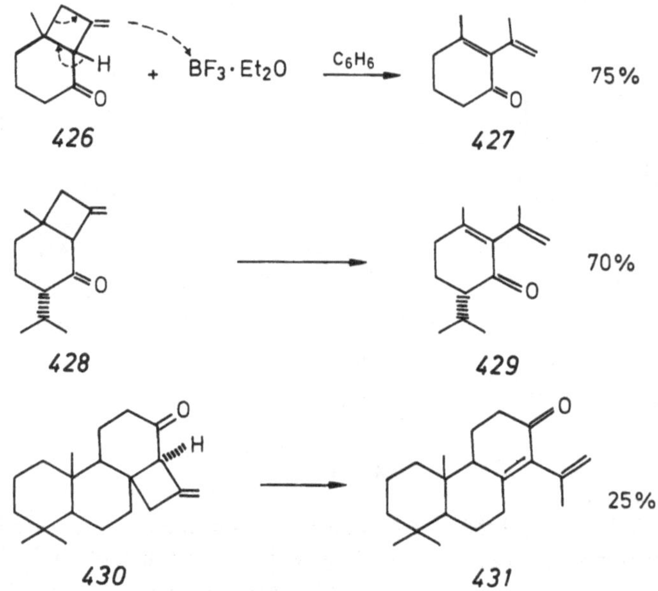

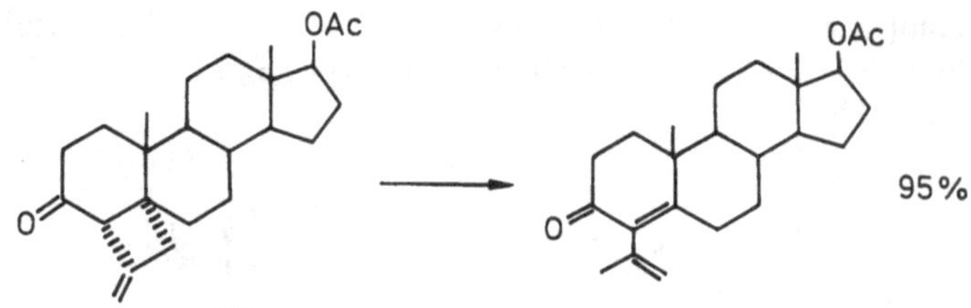

432 *433*

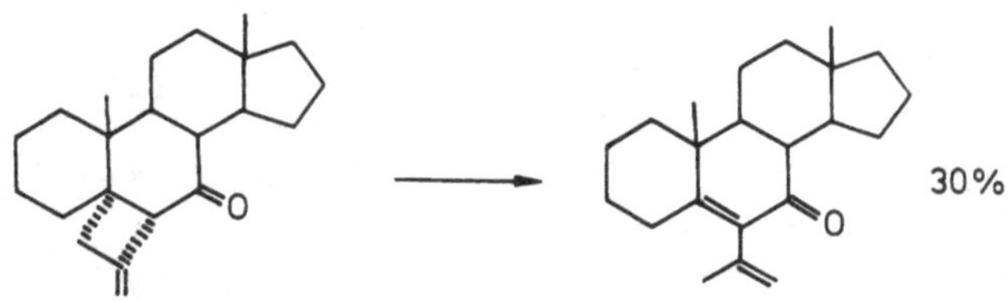

434 *435*

The potassium salt of compound (*436a*) readily underwent rearrangement at room temperature to provide a mixture of (*436a*) (25%), (*437a*) (67%) and (*437b*) (8%) [146]. Under the same conditions, (*436b*) was stable. In refluxing THF, or in the presence of 18-crown-6, (*436b*) isomerized to furnish (*437a*) and (*437b*) in a 6:1 ratio [146]. The isomerization of these substituted bicyclo[3.2.0]heptenes was believed to be converted, and the chemistry of these compounds has been reviewed [147].

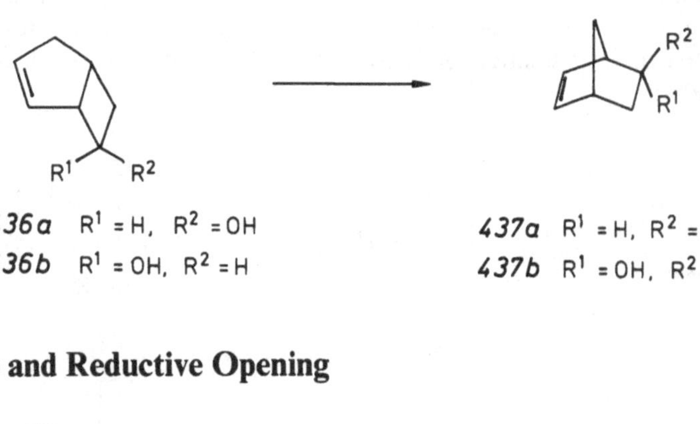

436a R^1 = H, R^2 = OH
436b R^1 = OH, R^2 = H

437a R^1 = H, R^2 = OH
437b R^1 = OH, R^2 = H

4 Oxidative and Reductive Opening

4.1 Oxidative Cleavage

1,2-Diols can be readily cleaved oxidatively to ketones by periodic acid (Eq. (17)).

$$\begin{matrix} R_2C{-}OH \\ | \\ R_2C{-}OH \end{matrix} + HIO_4 \rightleftharpoons \begin{matrix} R_2C{-}O \\ | \quad\; \rangle IO_4H_3 \\ R_2C{-}O \end{matrix} \longrightarrow 2R_2CO + HIO_3 \qquad \text{(Eq. 17)}$$

There are ample evidences which suggest that a cyclic periodate ester is the intermediate [148]. Cyclobutane-1,2-diols can also be cleaved oxidatively and this aspect has been used in organic synthesis. Thus, photocycloaddition of 1,2-*bis*(trimethylsiloxy)cyclobutene to (—)piperitone (*381*) gave the photoadduct (*438*). Desilylation of the compound (*438*) with fluoride ion, followed by oxidative cleavage of the 1,2-diol (*439*) with sodium periodate afforded the crystalline *cis*-bicyclo[4.4.0]decane-2,5,7-trione (*440*) [149].

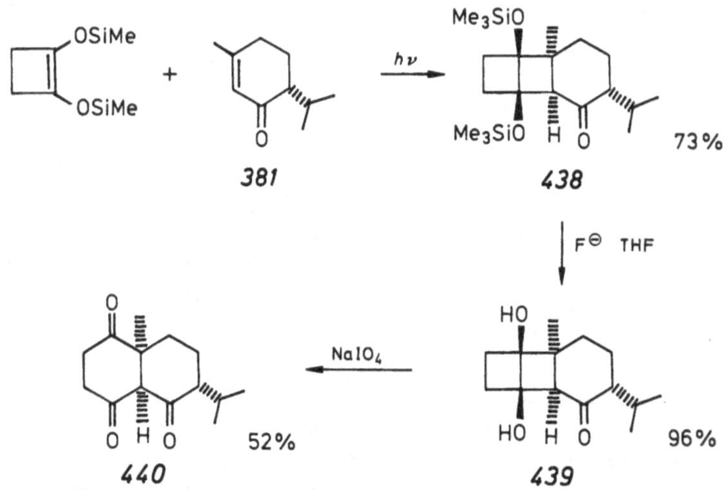

A similar result was reported by Vandewalle [150, 151]. Irradiation of 1,2-*bis*(trimethylsiloxy)cyclobutene and compounds (*441*) afforded compounds (*442*), which were reduced with $LiAlH_4$ to give the endo alcohol (*443*). Treatment of the alcohol (*443*) with methanol for 24 hours would bring about silylether cleavage as well as unexpected direct oxidation of (*443*) to the *cis*-hydrindanes (*444*) in 50–70% yield (Table 18) [150].

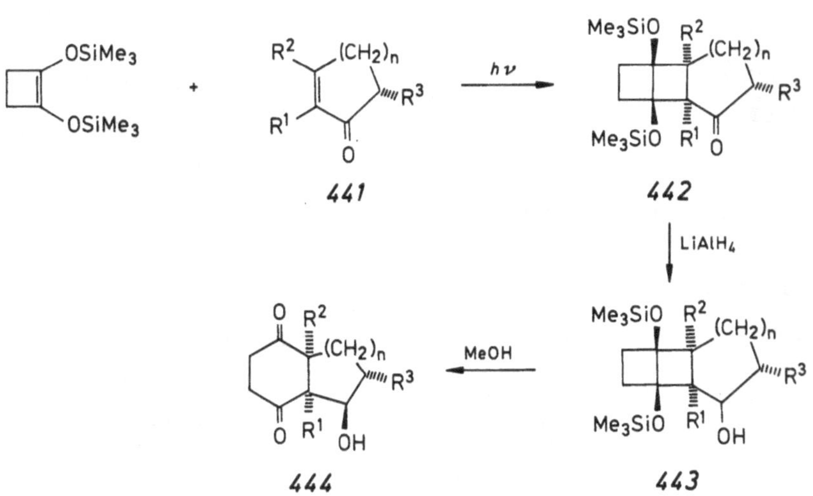

Photocycloaddition of 1,2-*bis*(trimethylsiloxy)cyclobutene and (*445*) gave the compound (*446*), which was reduced with $NaBH_4$, hydrolyzed and cleaved oxidatively to afford the epimeric product (*447*) [151].

Table 18

n	R¹	R²	R³
1	Me	H	H
1	H	Me	H
1	Me	Me	H
1	H	Me	ⁱPr
2	Me	H	H
2	H	Me	H
2	Me	Me	H
2	H	Me	ⁱPr

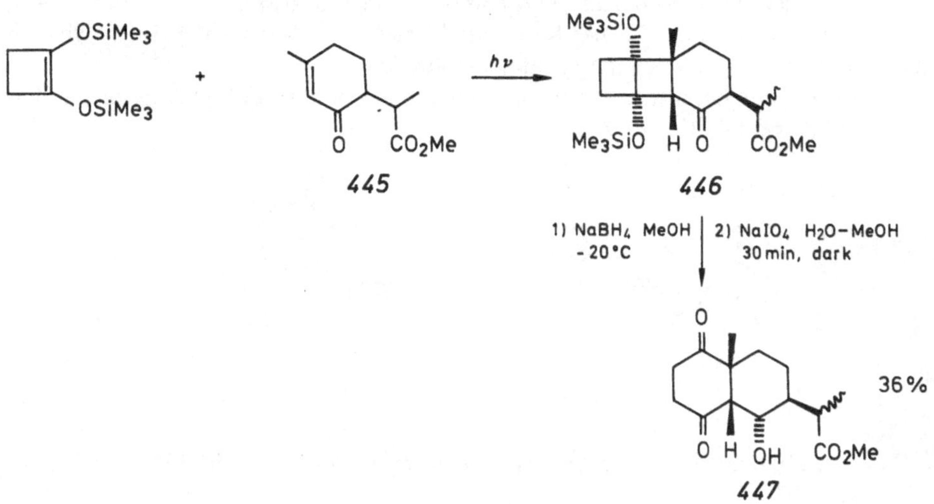

Cyclopentenones (*448a*) and (*448b*) underwent photocycloaddition with 1,2-*bis*[trimethylsiloxy]cyclopentene to give the epimeric (*449a*) and (*449b*) in 65% and

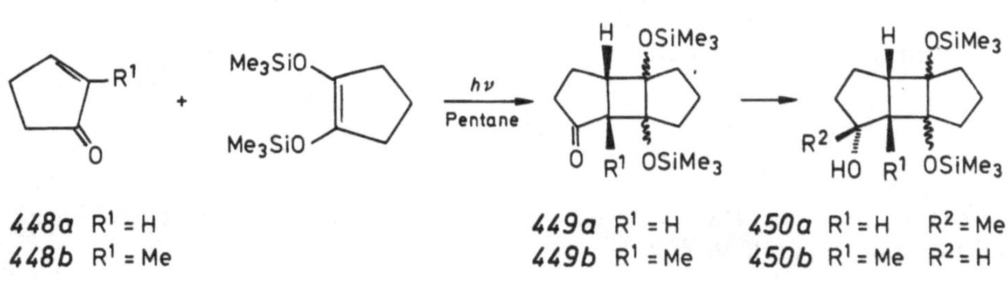

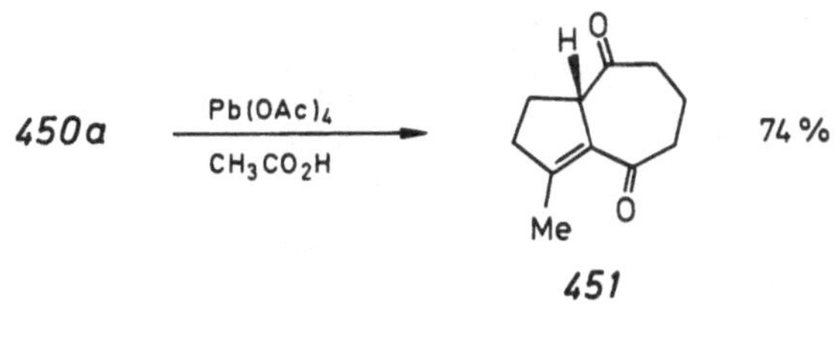

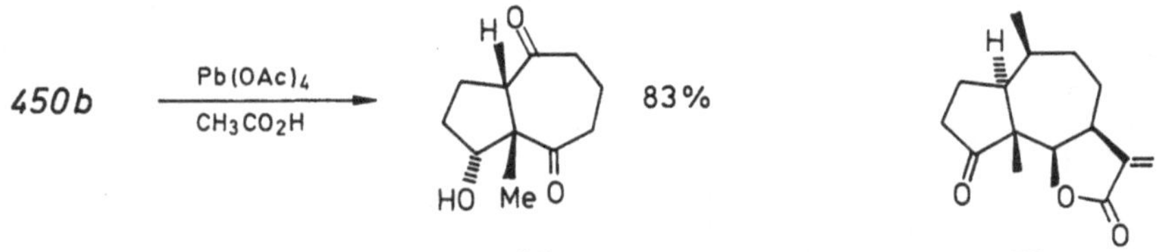

452 453

73% yield respectively. The ketones (*449a*) and (*449b*) were treated with methylmagnesium bromide and $LiAlH_4$ respectively to provide the alcohols (*450a*) and (*450b*) in 63% and 73% yield respectively [152].

Upon oxidation, (*450a*) and (*450b*) were cleaved to afford hydroazulenes (*451*) and (*452*). The alcohol (*452*) has been transformed to (±)-damsin (*453*) [153] which is a nonisoprenoid hydroazulenic pseudoguanianolide.

A similar route has also been applied to the total synthesis of (±)-compressanolide (*454*) and (±)-estafiatin (*455*) [154, 155].

454 455

The cyclobutane ring of (*456*) was cleaved by periodic acid in THF to give the diketone (*457*) [156].

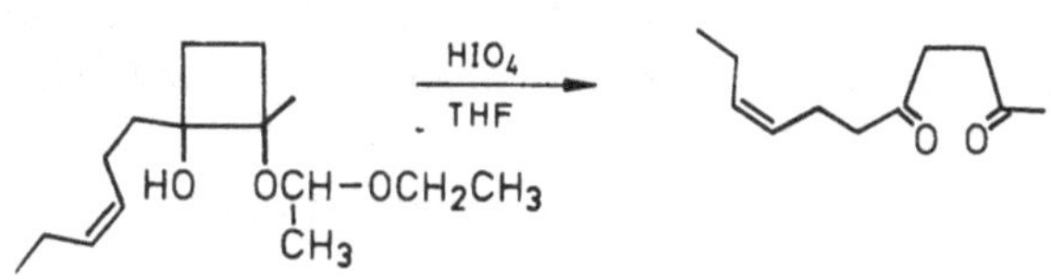

456 457

The cyclobutanols (*458*) underwent oxidative cleavage with chromic acid to provide diketones (*459*) (Table 19) [157].

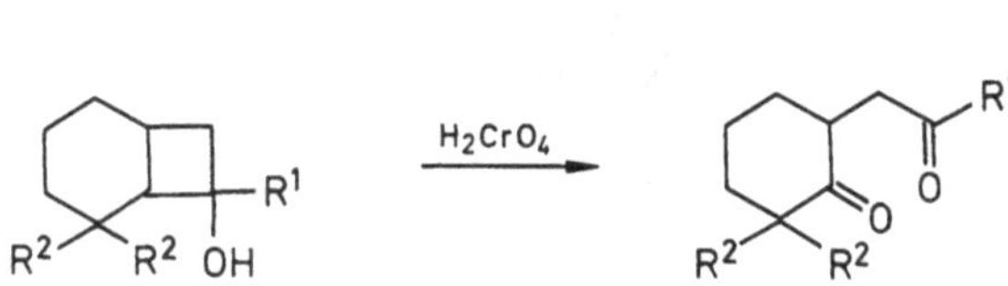

458 459

Table 19

R^1	R^2	yield of (*459*)
ⁿBu	H	82%
Me	Me	80%
Et	Me	78%
Me	H	85%

However, when there is a substituent on the bridgehead carbon, the product of the oxidation would be the keto-alcohols (*461*) and the ketone (*462*) (Table 20) [157].

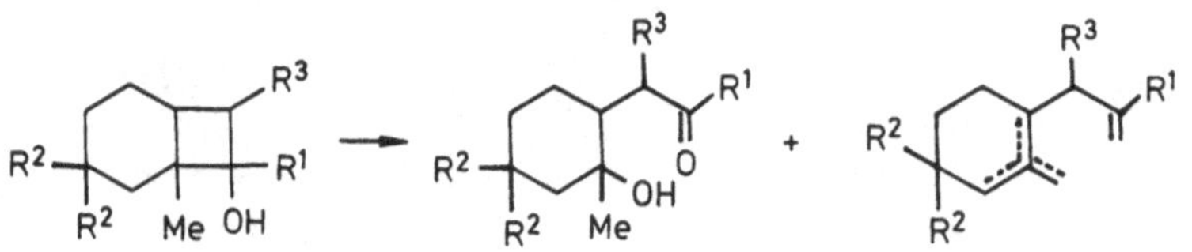

460 461 462

Table 20

R^1	R^2	R^3	yield of (*461*)	yield of (*462*)
Me	Me	H	50%	21%
ⁿBu	H	Me	48%	10%
Me	H	H	66%	—

The mechanism of the oxidation of tertiary cyclobutanols with Jones reagent is believed to involve the intermediate lactols (Eq. (18)) and the cleavage of the lactol to ketol and its subsequent oxidation to diketone when $R^1 = H$ [157].

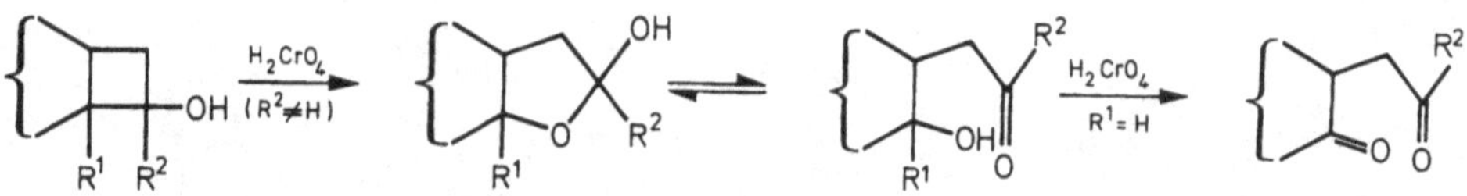

(Eq. 18)

Various oxidation conditions which could promote the ring cleavage of cyclobutanols have been briefly reviewed [158]. This work was extended to the cyclo-

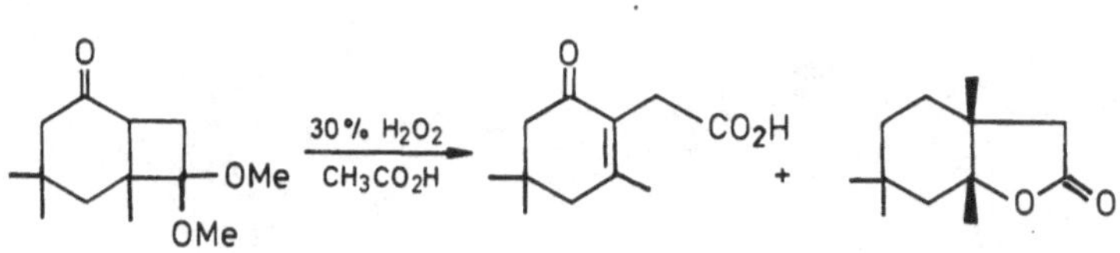

463 464 465

butanone dimethyl ketal (*463*), which was cleaved oxidatively to yield the β,γ-unsaturated acid (*464*) and the γ-lactone (*465*) in a ratio of 1:1 (60% yield) [159].

Cyclobutenes have been applied as latent functionality of 1,4-dicarbonyl systems. Photolysis of (*466*) gave a 1.5:1 mixture (60%) of (*467a*) and (*467b*), which were cleaved by ozone and subsequent reduction of the ozonides yielded the epimeric (*468*). Cyclization and dehydration process converted (*468*) to the furan (*469*) [160]. Furan (*469*) was converted to hibiscone C in a few steps [160].

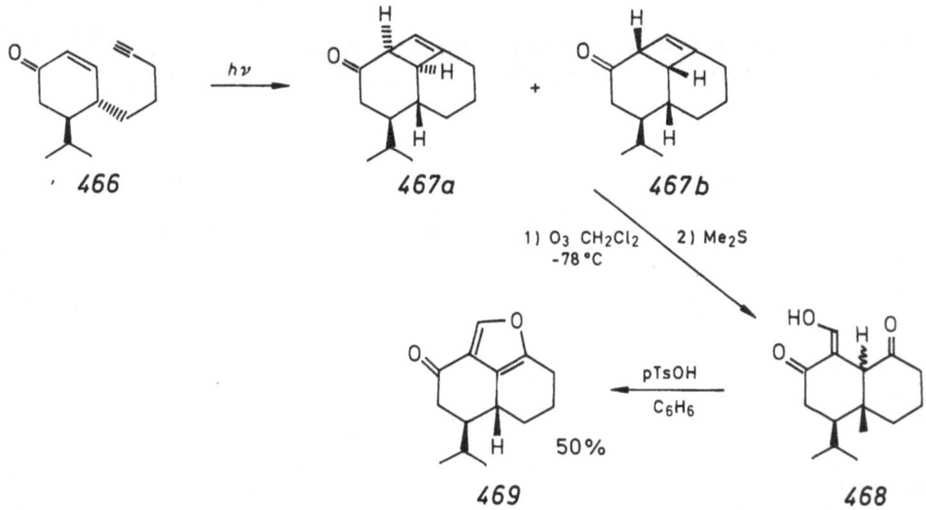

466 467a 467b 469 468

The ozonolysis of cyclobutene derivatives in the preparation of 1,4-diketones was also applied to the total synthesis of cyclopentanoid antibiotics [161,162]. The oxidative cleavage of (*470*) by ozone and reductive work-up yielded the diketone (*471*) in 73% yield. Diketone (*471*) underwent intramolecular aldol cyclization to give the key intermediate (*472*), which was used to synthesize (±)-xanthocidin [161,162], (±)-epi-xanthocidin [162], (±)-β-isoxanthocidin [161,162] as well as (±)-desdihydroxy-4,5-didehydroxanthocidin [162].

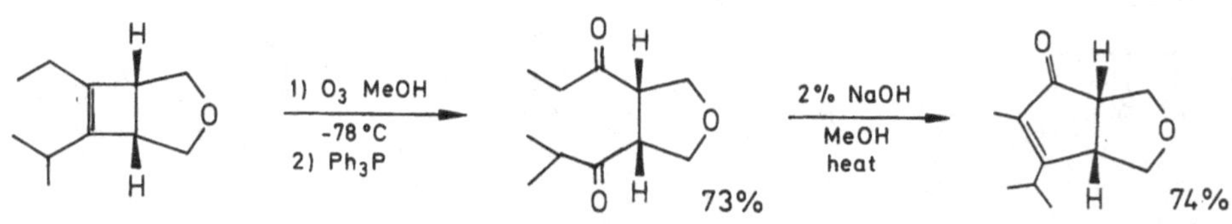

470 471 472

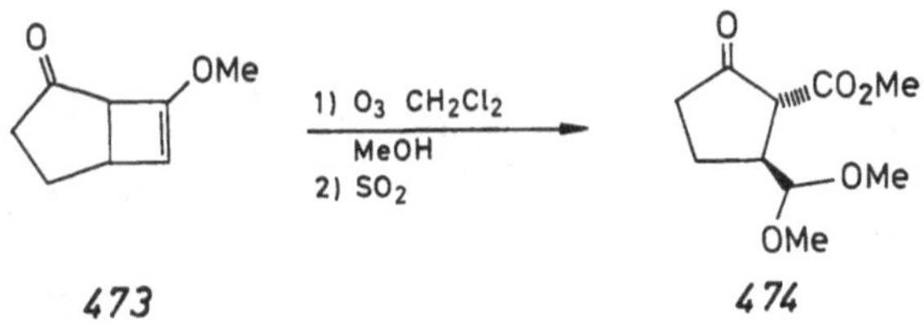

473 474

The methoxycylobutene (*473*) also underwent ozonolysis in the presence of MeOH and yielded (*474*) after reductive work-up with SO_2 [163]. The cyclopentane derivative (*474*) was finally converted to 11-deoxy-PGE_1 [163].

4.2 Reductive Cleavage

Cyclobutane derivatives can also be opened under reductive conditions. The reductive α,β-fragmentation of γ-halocyclobutylketones (Eq. (19)) has been studied

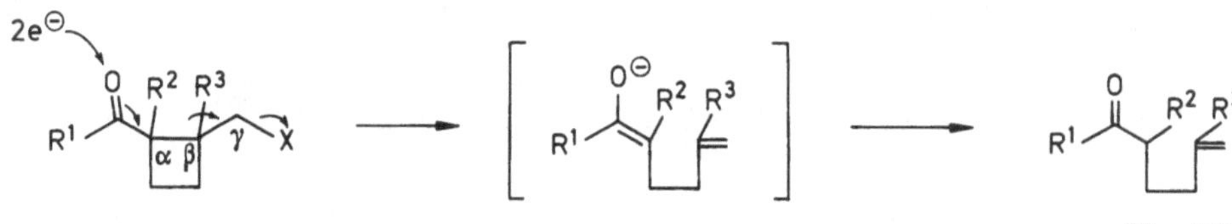

(Eq. 19)

extensively and developed to serve as a new approach to spiro systems [164]. Thus, irradiation of the compound (*475*) gave a mixture of photoadducts (*476*), (*477*) and other in an overall yield of 76%. The compound (*476*) was a mixture of two isolable epimers which were not identified stereochemically and were converted to (*478*) on reductive cleavage (Li/liq. NH_3, THF, −33 °C) to furnish the spiroketone (*478*) in 66–79% yield. The compound (*478*) underwent further transformation steps to give (±)-α-acoradiene (*479*) [165].

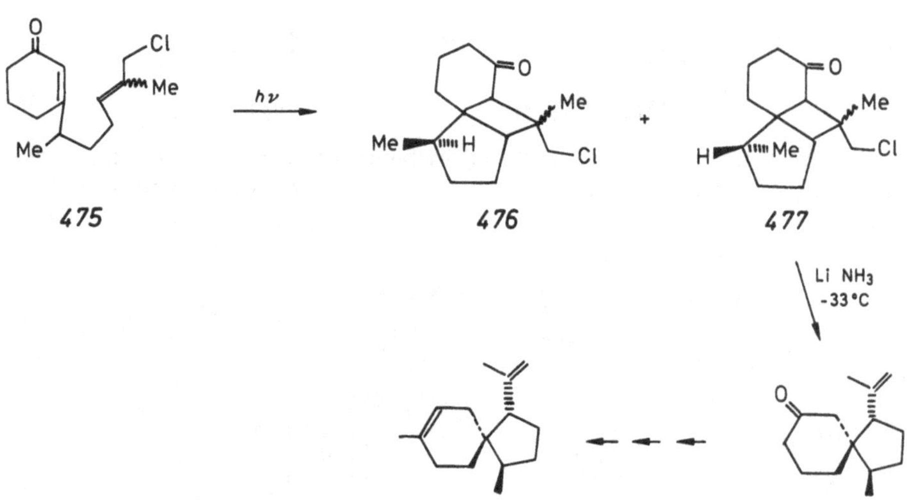

475 476 477 479 478

1,4-dicarbonyl compounds can be cleaved at the 3,4 carbon—carbon bond under reduction conditions. The sequence (*480*)→(*481*)→(*482*) reveals a new strategy for building medium ring compounds by beginning with an intramolecular photochemical [2 + 2]cycloaddition of fused α,β-unsaturated γ-lactones (*480*) bearing alkenyl side chains at the γ-position, and was followed by a reductive cleavage of the cyclobutane ring that formed. The application of this sequence was shown as follows [166]. Irradiation of the lactone (*483*) in the presence of acetone afforded the compound (*484*). The compound (*484*) was subject to saponification, oxidation and esterification to give the ketoester (*485*). Finally, (*485*) was converted to the compound (*486*) by reductive cleavage of the cyclobutane ring [166].

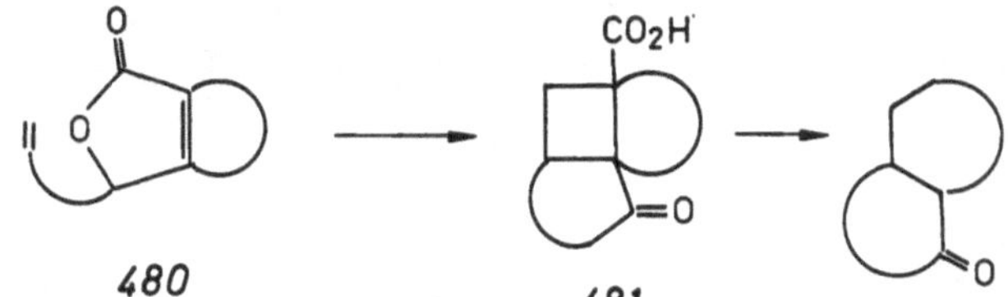

480 *481* *482*

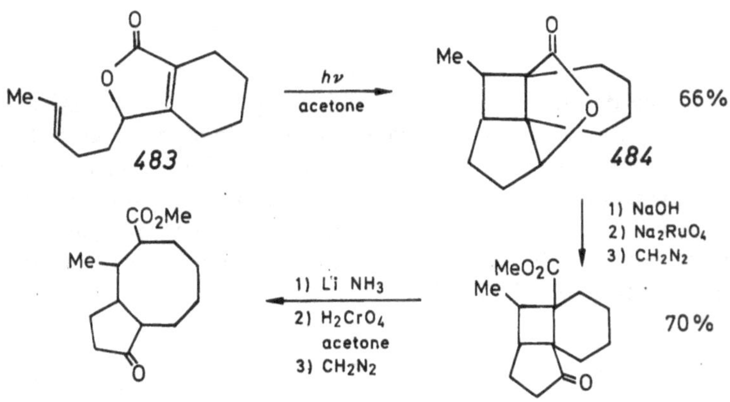

483 *484* *486* *485*

On the other hand, irradiation of the allenylbutenolide (*487*) in *p*-xylene gave the photoadduct (*488*), which was also subject to saponification, oxidation and esterifica-

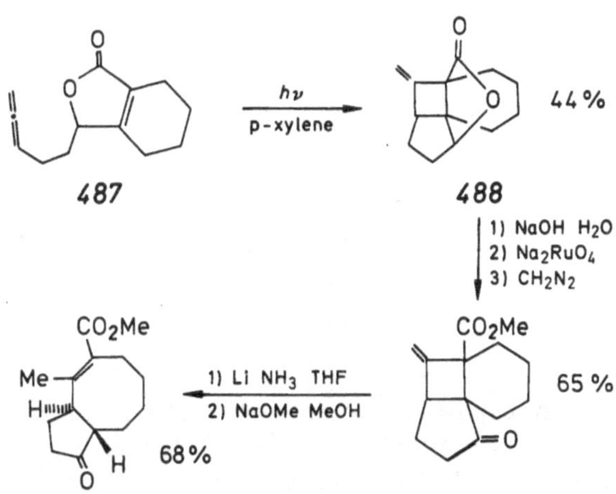

487 *488* *490* *489*

tion in a similar manner, providing the keto-ester (*489*). Reduction of compound (*489*) with lithium in liq. NH_3 and subsequent treatment with NaOMe in MeOH gave the ester (*490*) [166].

The cyclobutane dione (*491*) was converted quantitatively to the diketone (*492*) by treatment with zinc in HOAC. The reaction involves also the reductive cleavage of 1,4-dicarbonyl compounds [167,168].

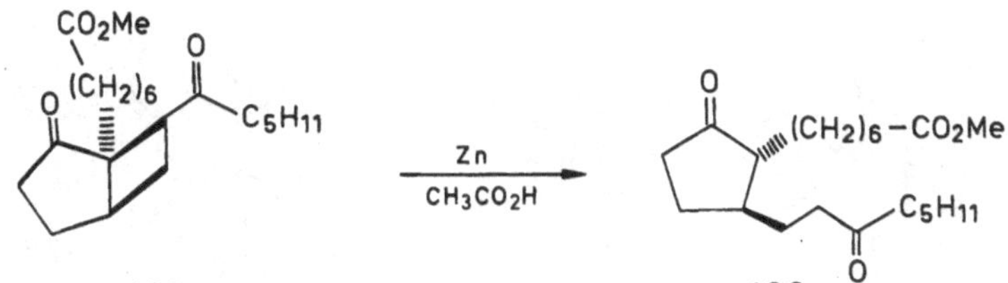

491 *492*

Reduction of *cis*-1,2-divinylcyclobutane (*493a*) with alkali metals such as Na in liq. NH_3 resulted in ring cleavage to give *cis,trans*-2,6-octadiene (*494a*) in 69% yield [169]. Similarly, the dimethyl derivative (*493b*) gave (*494b*) in 72% yield [169]. The first step in the reduction of (*493a*) and (*493b*) probably involves the formation of radical anions, and is followed by the cleavage of the cyclobutane rings and the protonation of allyl anions [169].

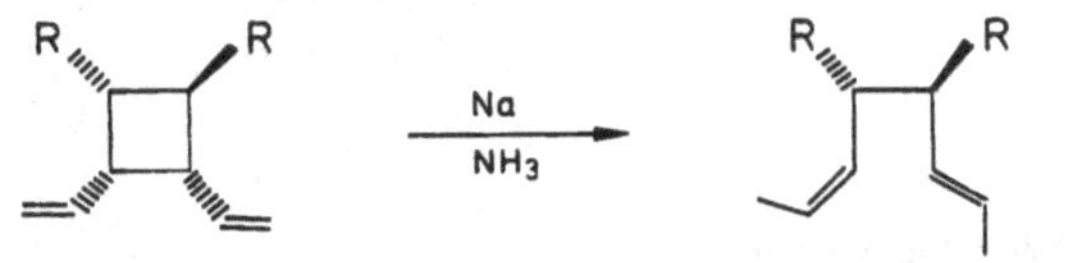

493a R = H *494a* R = H
493b R = Me *494b* R = Me

Reductive cleavage procedure could be applied to (*495*), which upon reduction by several conditions, provided different products (*496*) (Table 21) [170]. Similarly, (*498*) and (±)-10-epijunenol (*500*) could be obtained from the reduction of (*497*) and (*499*) [170].

Table 21

(*493*)	Condition	(*494*)	yield of (*494*)
R = CO_2Me	Li/liq. NH_3/—33 °C Et_2O/^tBuOH	$R^1 = CH_2OH, R^2 = H$	87%
R = CO_2Me	1) Li/liq. NH_3/THF aniline 2) NH_4Cl	$R^1 = CO_2Me, R^2 = H$	79%
R = CO_2Me	1) Li/liq. NH_3/THF aniline 2) MeI	$R^1 = CO_2Me, R^2 = Me$	76%
R=CH_2–P(=O)$(OEt)_2$	Li/liq. NH_3	$R^1R^2 = CH_2$	

495 *496*

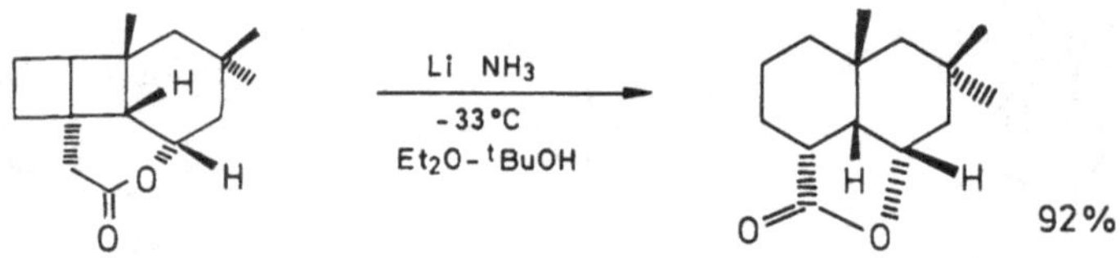

497 *498*

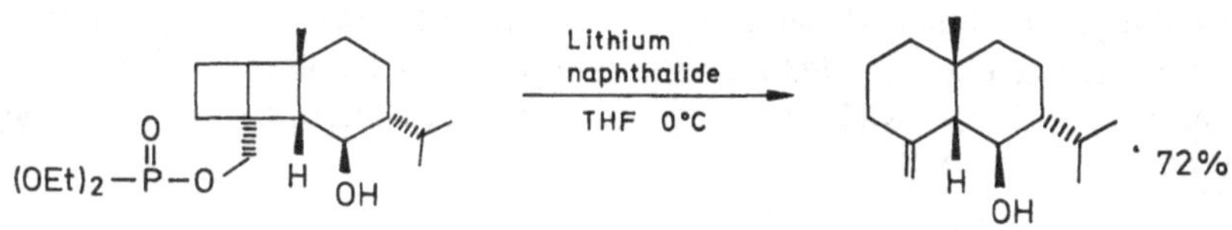

499 *500*

5 Miscellaneous Ring Openings

The compound (*501*) was converted to $\Delta^{2(3)}$-13-norilluden-7-one (*502*) by treatment with AgOAc in HOAc (120 °C, 7 hrs) [171]. This transformation was proposed to be a biogenetic type reaction [171].

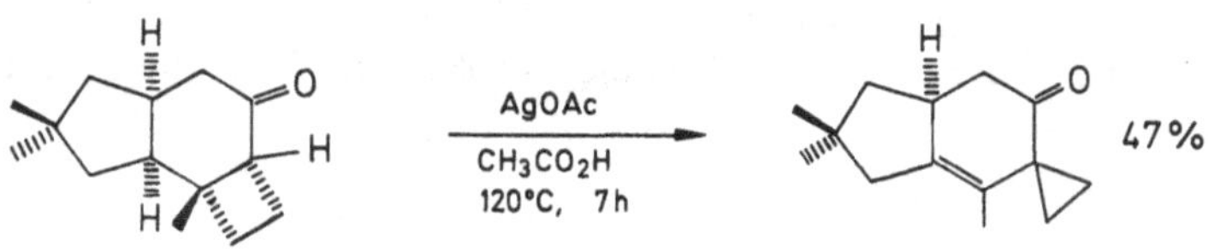

501 *502*

The migration of the trimethylsilyl group would help to open the four-membered ring of (*503*), and as a result, the dichlorotrimethylsilylenol ether (*504*) was

503 *504*

isolated [172]. This process was proposed to be catalyzed by $ZnCl_2$ which was present in the reaction mixture [172]. Similar observation was also reported recently [12].

Photochemical [2 + 2]cycloaddition of (*505*) to (*506*) provided a mixture of head to tail and head to head isomers (*507*) and (*508*) which was not separated but was subjected to addition of methyllithium. The tertiary alcohols (*509*) and (*510*) obtained were readily separated on trituration and the desired compound (*509*) was converted

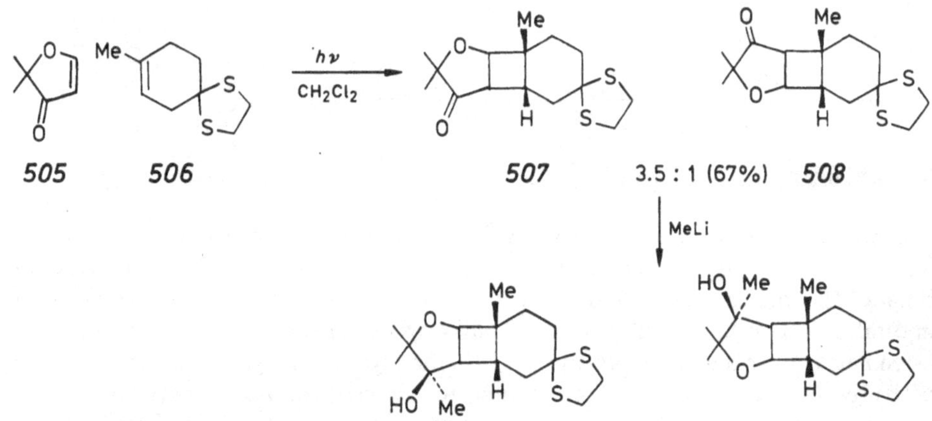

505 506 507 508 509 510

to the nitrile ester (*511*) by reaction with nitrosyl chloride. The nitrile ester (*511*) underwent facile fragmentation on irradiation to provide the keto-aldehyde (*512*), which, upon treatment with acid, cyclized to give octalone (*513*) [173]. This process was applied extensively for the synthesis of substituted cyclohexenones (*516*) from (*514*) via the dicarbonyl compounds (*515*) (Table 22) [174].

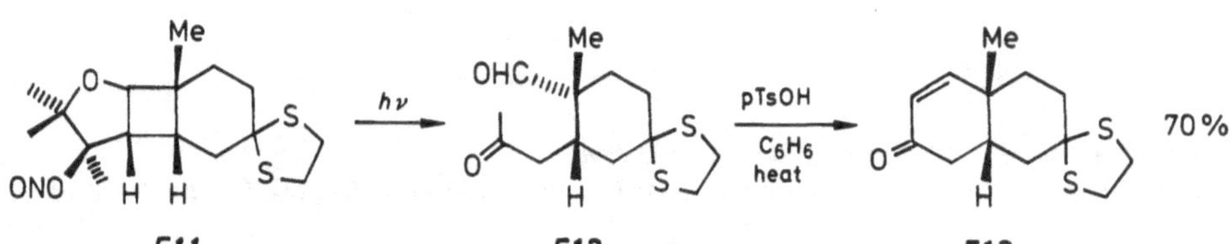

511 512 513

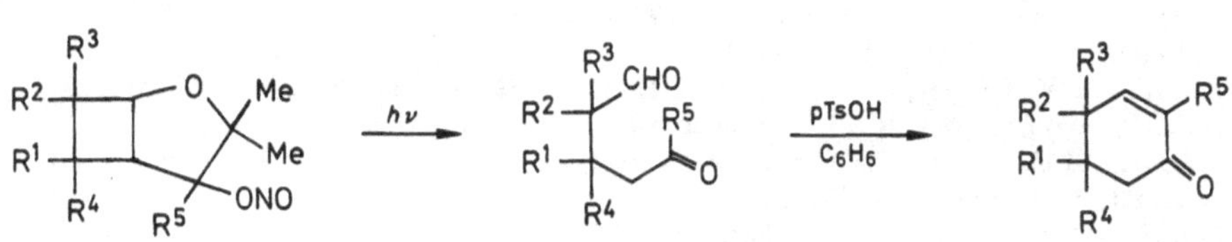

514 515 516

Table 22

R^1	R^2	R^3	R^4	R^5	yield of (*516*) from (*514*)
H	Me	Me	H	Me	93%
H	Me	Me	H	^{n}Bu	85%
H	Me	CH_2Bu	H	Me	93%
Me	Me	Me	H	Me	74%
$-(CH_2)_4-$		H	H	Me	80%
$-(CH_2)_3-$		H	H	Me	88%
$-(CH_2)_4-$		Me	H	Me	92%
$-(CH_2)_3-$		Me	H	Me	98%
H	$-(CH_2)_4-$		H	Me	78%

6 Summary and Outlook

Strained molecules such as cyclopropanes and cyclobutanes have emerged as important intermediates in organic synthesis. We have already demonstrated here that cyclobutane derivatives can indeed serve as starting materials for the synthesis of natural as well as unnatural products. Unlike cyclopropanes, which can be prepared asymmetrically in a number of ways [175–182], the asymmetric synthesis of cyclobutane derivative has received less attention, and, to our best knowledge, very few reports were recorded recently [183]. Obviously, the ready availability of chiral cyclobutane derivatives would greatly enhance their usefulness in the enantioselective synthesis of natural products. The overcome of this last hurdle would allow cyclobutane derivatives to play an even more important role in synthetic organic chemistry.

7 Acknowledgement

We wish to thank Miss Sau Ling Chim for typing the manuscript and Mr. Kim Fah Liew for drawing all the figures.

8 References

1. Lukina, M. Yu.: Russ. Chem. Rev. *32*, 635 (1963)
2. Ginsburg, D. (Ed.): Alicyclic Compounds: Int. Rev. of Science, Organic Chemistry, series 2, vol. 5, Butterworths, London 1976, pp. 83–87
3. Wilson, A., Goldhamer, D.: J. Chem. Educ. *40*, 599 (1963)
4. Wilson, A., Goldhamer, D.: ibid. *40*, 504 (1963)
5. de Mayo, P., Takeshita, H., Satter, A. B. M. A.: Proc. Chem. Soc. *1962*, 119
6. de Mayo, P., Takeshita, H.: Can J. Chem. *41*, 440 (1963)
7. de Mayo, P.: Acc. Chem. Res. *4*, 49 (1971)
8. Baldwin, S. W., Gawley, R. E., Doll, R. J., Leung, K. H.: J. Org. Chem. *40*, 1865 (1975); Baldwin, S. W., Gawley, R. E.: Tetrahedron Lett. *1975*, 3969
9. Pearlman, B. A.: J. Am. Chem. Soc. *101*, 6398, (1979)
10. Pearlman, B. A.: ibid. *101*, 6404 (1979)
11. Liu, H.-J., Chan, W. H.: Can J. Chem. *58*, 2196 (1980)
12. Brady, W. T., Lloyd, R. M.: J. Org. Chem. *46*, 1322 (1981)

13. Liu, H.-J., Dieck-Abularach, T.: Tetrahedron Lett. *23*, 295 (1982)
14. Ayer, W. A., Ward, D. E., Browne, L. M., Delbaere, L. T. J., Hoyano, Y.: Can. J. Chem. *59*, 2665 (1981)
15. Goldstein, S., Vannes, P., Houge, C., Frisque-Hesbain, A. M., Wiaux-Zamar, C., Ghosez, L., Germain, G., Declercq, J. P., Van Meerssche, M., Arrieta, J. M.: J. Am. Chem. Soc. *103*, 4616 (1981)
16. Lange, G. L., de Mayo, P.: J. Chem. Soc. Chem. Commun. *1967*, 704
17. Oppolzer, W. Bird, T. G. C.: Helv. Chim. Acta *62*, 1199 (1979)
18. Begley, M. J., Mellor, M., Pattenden, G.: J. Chem. Soc. Chem. Commun. *1979*, 235
19. Birch, A. M., Pattenden, G.: ibid. *1980*, 1195
20. Birch, A. M., Pattenden, G.: J. Chem. Soc. Perkin Trans. 1 *1983*, 1913
21. Pauw, J. E., Weedon, A. C.: Tetrahedron Lett. *23* 5485 (1982)
22. Begley, M. J., Mellor, M., Pattenden, G.: J. Chem. Soc. Perkin Trans. 1 *1983*, 1905
23. Büchi, G., Carlson, J. A., Powell, Jr., J. E., Tietze, L.-F.: J. Am. Chem. Soc. *95*, 540 (1973)
24. Tietze, L.-F.: Angew. Chem. *85*, 763 (1973); Angew. Chem. Int. Ed. Engl. *12*, 757 (1973); Chem. Ber. *107*, 2499 (1974)
25. Partridge, J. J., Chadha, N. K., Uskoković, M. R.: J. Am. Chem. Soc. *95*, 532 (1973)
26. Baldwin, S. W., Crimmins, M. T., Cheek, V. I.: Synthesis *1978*, 210
27. Ficini, J., d'Angelo, J., Eman, A., Touzin, A. M.: Tetrahedron Lett. *1976*, 683
28. Ogino, T., Yamada, K., Isogai, K.: ibid. *1977*, 2445
29. Van Audenhove, M., Termont, D., De Keukeleire, D., Vandewalle, M., Claeys, M.: ibid. *1978*, 2057
30. Kelly, R. B., Zamecnik, J., Beckett, B. A.: Can. J. Chem. *50*, 3455 (1972)
31. Kelly, R. B., Eber, J., Hung, H. K.: ibid. *51*, 2534 (1973)
32. Kelly, R. B., Harley, M. L., Alward, S. J., Rej, R. N., Gowda, G., Mukhopadhyay, A., Manchand, P. S.: ibid. *61*, 269 (1983)
33. Wiesner, K., Ho, P.-T., Liu, W.-C., Shanbhag, M. N.: ibid. *53*, 2140 (1975)
34. Wiesner, K.: Pure and Applied Chem. *41*, 93 (1975)
35. Guthrie, R. W., Valenta, Z., Wiesner, K.: Tetrahedron Lett. *1966*, 4645
36. Wiesner, K., Sanchez, I. H., Atwal, K. S., Lee, S. F.: Can. J. Chem. *55*, 1091 (1977)
37. Wiesner, K., Musil, V., Wiesner, K. J.: Tetrahedron Lett. *1968*, 5643
38. For reviews, see Grob, C. A.: Angew. Chem. *81*, 543 (1969); Angew. Chem. Int. Ed. Engl. *8*, 535 (1969); Grob, C. A., Schiess, P. W.: Angew. Chem. *79*, 1 (1967); Angew. Chem. Int. Ed. Engl. *6*, 1 (1967)
39. Oppolzer, W.: Acc. Chem. Res. *15*, 135 (1982)
40. Challand, B. D., Hikino, H., Kornis, G., Lange, G., de Mayo, P.: J. Org. Chem. *34*, 794 (1969)
41. Tietze, L.-F., Reichert, U.: Angew. Chem. *92*, 832 (1980); Angew. Chem. Int. Ed. Engl. *19*, 830 (1980)
42. Oppolzer, W., Wylie, R. D.: Helv. Chim. Acta *63*, 1198 (1980)
43. Tatsuta, K., Akimoto, K., Kinoshita, M.: J. Am. Chem. Soc. *101*, 6116 (1979)
44. Tatsuta, K., Akimoto, K., Kinoshita, M.: Tetrahedron *37* 4365 (1981)
45. Barker, A. J., Begley, M. J., Mellor, M., Otieno, D. A., Pattenden, G.: J. Chem. Soc. Perkin Trans. 1 *1983*, 1893
46. Barker, A. J., Pattenden, G.: J. Chem. Soc. Perkin Trans. 1 *1983*, 1901
47. Liu, H.-J., Lee, S. P.: Tetrahedron Lett. *1977*, 3699
48. Liu, H.-J., Valenta, Z., Wilson, J. S., Yu, T. T. J.: Can. J. Chem. *47*, 509 (1969)
49. Brady, W. T.: Tetrahedron *37*, 2949 (1981); Synthesis *1971*, 415
50. Sauer, J. C.: J. Am. Chem. Soc. *69*, 2444 (1947); Brady, W. T., Scherubel, G. A.: J. Org. Chem. *39*, 3790 (1974)
51. Staudinger, H.: Chem. Ber. *38*, 1735 (1905); McCarney, C. C., Ward, R. S.: J. Chem. Soc., Perkin Trans. 1 *1975*, 1600
52. Greene, A. E., Luche, M.-J., Deprés, J.-P.: J. Am. Chem. Soc. *105*, 2435 (1983)
53. Au-Yeung, B. W., Fleming, I.: J. Chem. Soc. Chem. Commun. *1977*, 81
54. Fleming, I., Au-Yeung, B. W.: Tetrahedron *37* (S 1), 13 (1981)
55. Liu, H.-J., Ogino, T.: Tetrahedron Lett. *1973*, 4937

56. Annis, G. D., Paquette, L. A.: J. Am. Chem. Soc. *104*, 4504 (1982)
57. Trost, B. M.: Acc. Chem. Res. *7*, 85 (1974); Fortschr. Chem. Forsch. *41*, 1 (1973)
58. Trost, B. M., Latimer, L. H.: J. Org. Chem. *43*, 1031 (1978)
59. Halazy, S., Krief, A.: J. Chem. Soc. Chem. Commun. *1982*, 1200
60. Halazy, S., Zutterman, F., Krief, A.: Tetrahedron Lett. *23*, 4385 (1982)
61. Gadwood, R. C.: J. Org. Chem. *48*, 2098 (1983)
62. Rühlmann, K.: Synthesis *1971*, 236; cf. Rühlmann, K., Segfluth, H., Beck, H.: Chem. Ber. *100*, 3820 (1967); Bloomfield, J. J., Owsley, D. C., Nelke, J. M.: Org. React. *23*, 259 (1976)
63. Nakamura, E., Kuwajima, I.: J. Am. Chem. Soc. *99*, 961 (1977)
64. Nishiguchi, I., Hirashima, T., Shono, T., Sasaki, M.: Chem. Lett. *1981*, 551
65. For reviews, see Lee, J. B., Uff, B. C.: Q. Rev. Chem. Soc. *21*, 429 (1967); Hassall, C. H.: Org. React. *9*, 73 (1957)
66. Au-Yeung, B. W., Fleming, I.: J. Chem. Soc. Chem. Commun. *1977*, 79
67. Grieco, P. A., Oguri, T., Wang, C.-L. J., Williams, E.: J. Org. Chem. *42*, 4113 (1977)
68. Grieco, P. A., Oguri, T., Gilman, S., DeTitta, G. T.: J. Am. Chem. Soc. *100*, 1616 (1978)
69. Smith, III, A. B., Richmond, R. E.: J. Org. Chem. *46*, 4814 (1981); J. Am. Chem. Soc. *105*, 575 (1983)
70. Still, W. C., Murata, S., Revial, G., Yoshihara, K.: J. Am. Chem. Soc. *105*, 625 (1983)
71. Jeffs, P. W., Molina, G., Cass, M. W., Cortese, N. A.: J. Org. Chem. *47*, 3871 (1982)
72. Deprés, J.-P., Greene, A. E., Crabbé, P.: Tetrahedron *37*, 621 (1981)
73. Ho, P.-T., Lee, S. F., Chang, D., Wiesner, K.: Experientia *27*, 1377 (1971)
74. Michel, P., O'Donnell, M., Biname, R., Hesbain-Frisque, A. M., Ghosez, L., Declercq, J. P., Germain, G., Arte, E., Van Meerssche, M.: Tetrahedron Lett. *21*, 2577 (1980)
75. Piers, E., Abeysekera, B. F., Herbert, D. J., Suckling, I. D.: J. Chem. Soc. Chem. Commun. *1982*, 404
76. Helmlinger, D., de Mayo, P., Nye, M., Westfelt, L., Yeats, R. B.: Tetrahedron Lett. *1970*, 349
77. Ziegler, F. E., Kloek, J. A.: Tetrahedron *33*, 373 (1977)
78. Trost, B. M., Rigby, J. H.: J. Org. Chem. *41*, 3217 (1976)
79. Terlinden, R., Boland, W., Jaenicke, L.: Helv. Chim. Acta *66*, 466 (1983)
80. Miller, R. D., McKean, D. R.: Tetrahedron Lett. *21*, 2639 (1980)
81. Breslow, R., in de Mayo, P. (Ed.): Molecular Rearrangments, Interscience, New York 1963, pp. 233–294
82. Matz, J. R., Cohen, T.: Tetrahedron Lett. *22*, 2459 (1981)
83. Barton, T. J., Kippenhan, Jr., R. C., Nelson, A. J.: J. Am. Chem. Soc. *96*, 2272 (1974)
84. Stevens, H. C., Reich, D. A., Brandt, D. R., Fountain, K. R., Gaughan, E. J.: ibid. *87*, 5257 (1967)
85. Bartlett, P. D., Ando, T.: ibid. *92*, 7518 (1970)
86. Asao, T., Machiguchi, T., Kitamura, T., Kitahara, Y.: J. Chem. Soc. Chem. Commun. *1970*, 89
87. Tanaka, K., Yoshihoshi, A.: Tetrahedron *27*, 4889 (1971)
88. Turner, R. W., Seden, T.: J. Chem. Soc. Chem. Commun. *1966*, 399
89. Wiesner, K., Uyeo, S., Philipp, A., Valenta, Z.: Tetrahedron Lett. *1968*, 6279
90. Guthrie, R. W., Henry, W. A., Immer, H., Wong, C. M., Valenta, Z., Wiesner, K.: Coll. Czech. Chem. Commun. *31*, 602 (1966)
91. Do Khac Manh Dur, Fetizon, M., Lazare, S.: J. Chem. Soc. Chem. Commun. *1975*, 282
92. White, J. D., Matsui, T., Thomas, J. A.: J. Org. Chem. *46*, 3376 (1981)
93. Takeda, K., Shimono, Y., Yoshii, E.: J. Am. Chem. Soc. *105*, 563 (1983)
94. Ochiai, M., Arimoto, M., Fujita, E.: J. Chem. Soc. Chem. Commun. *1981*, 460
95. Cargill, R. L., Dalton, J. R., O'Connor, S., Michels, D. G.: Tetrahedron Lett. *1978*, 4465
96. Smith, III, A. B., Jerris, P. J.: J. Am. Chem. Soc. *103*, 194 (1981)
97. Pirrung, M. C.: ibid. *101*, 7130 (1979); ibid. *103*, 82 (1981)
98. Tobe, Y., Ueda, Y., Nishikawa, H., Odaira, Y.: J. Org. Chem. *46*, 5009 (1981)
99. Baudouy, R., Crabbé, P., Greene, A. E., Le Drian, C., Orr, A. F.: Tetrahedron Lett. *1977*, 2973

100. Becker, D., Harel, Z., Nagler, M., Gillon, A.: J. Org. Chem. *47*, 3297 (1982)
101. Kirmse, W., Streu, J.: Synthesis *1983*, 994
102. Benner, J. P., Gill, G. B., Parrott, S. J., Wallace, B.: J. Chem. Soc. Perkin Trans. 1 *1984*, 331
103. Francisco, C. G., Freire, R., Hernández, R., Melián, D., Salazar, J. A., Suárez, E.: J. Chem. Soc. Perkin Trans. 1 *1984*, 459
104. Taber, D. F.: Intramolecular Diels-Alder and Alder Ene Reactions, Springer-Verlag, Berlin 1984;
Fallis, A. G.: Can J. Chem. *62*, 183 (1984).
Brieger, G., Bennett, J. N.: Chem. Rev. *80*, 63 (1980)
105. Ishihara, A., Kimura, R., Yamada, S., Sakamura, S.: J. Am. Chem. Soc. *102*, 6353 (1980)
106. Fukumoto, K., Chihiro, M., Shiratori, Y., Ihara, M., Kametani, T., Honda, T.: Tetrahedron Lett. *23*, 2973 (1982);
McCullough, J. J.: Acc. Chem. Res. *13*, 270 (1980)
107. Kametani, T., Fukumoto, K.: Synthesis *1976*, 319;
Kametani. T.. Fukumoto, K.: Heterocycles *8*, 519 (1977);
Kametani, T.: Pure and Applied Chem. *51*, 747 (1979);
Oppolzer, W.: Angew. Chem. Int. Ed. Engl. *16*, 10 (1977);
Oppolzer, W.: Synthesis *1978*, 793;
Oppolzer, W.: Heterocycles *14*, 1615 (1980);
Vollhardt, K. P. C.: Acc. Chem. Res. *10*, 1 (1977);
Vollhardt, K. P. C.: Ann N.Y. Acad. Sci. *333*, 241 (1980);
Magnus, P., Gallagher, T., Brown, P., Pappalardo, P.: Acc. Chem. Res. *17*, 35 (1984)
108. Jung, M. E., Halweg, K. M.: Tetrahedron Lett. *22*, 2735 (1981)
109. Kametani, T., Tsubuki, M., Nemoto, H., Suzuki, K.: J. Am. Chem. Soc. *103*, 1256 (1981)
110. Boeckman, Jr., R. K., Delton, M. H., Nagasaka, T., Watanabe, T.: J. Org. Chem. *42*, 2946 (1977)
111. Oppolzer, W., Keller, K.: J. Am. Chem. Soc. *93*, 3836 (1971)
112. Funk, R. L., Vollhardt, K. P. C.: ibid. *98*, 6755 (1976)
113. Kametani, T., Honda, T., Matsumoto, H., Fukumoto, K.: J. Chem. Soc. Perkin Trans. 1 *1981*, 1383
114. Kametani, T., Suzuki, K., Nemoto, H.: J. Chem. Soc. Chem. Commun. *1979*, 1127
115. Grieco, P. A., Takigawa, T., Schillinger, W. J.: J. Org. Chem. *45*, 2247 (1980)
116. Kametani, T., Matsumoto, H., Nemoto, H., Fukumoto, K.: J. Am. Chem. Soc. *100*, 6218 (1978)
117. Nicolaou, K. C., Barnette, W. E., Ma, P.: J. Org. Chem. *45*, 1463 (1980)
118. Funk, R. L., Vollhardt, K. P. C.: J. Am. Chem. Soc. *102*, 5253 (1980)
119. Naito, T., Kaneko, C.: Tetrahedron Lett. *22*, 2671 (1981)
120. Rhoads, S. J., in de Mayo, P. (Ed.): Molecular Rearrangments, Interscience, New York 1963, pp. 655–706;
Rhoads, S. J., Raulins, N. R.: Org. React. *22*, 1 (1975);
Vogel, E.: Angew. Chem. *75*, 829 (1962); Angew. Chem. Int. Ed. Engl. *2*, 1 (1963); von E. Doering, W., Roth, W. R.: Angew. Chem. *75*, 27 (1963); Angew. Chem. Int. Ed. Engl. *2*, 115 (1963)
121. For a review, see Marvell, E. N., Whalley, W., in Patai, S. (Ed.): The Chemistry of the Hydroxy Group, pt. 2, Interscience, New York 1971, pp. 738–743
122. Schreiber, S. L., Santini, C.: Tetrahedron Lett. *22*, 4651 (1981)
123. Jung, M. E., Hatfield, G. L.: ibid. *24*, 2931 (1983)
124. Paquette, L. A., Andrews, D. R., Springer, J. P.: J. Org. Chem. *48*, 1147 (1983)
125. Danheiser, R. L., Gee, S. K., Sard, H.: J. Am. Chem. Soc. *104*, 7670 (1982)
126. Grubbs, R. H.: Prog. Inorg. Chem. *24*, 1 (1978);
Katz, T. J.: Adv. Organomet. Chem. *16*, 283 (1977);
Basset, J. M., Leconte, M.: CHEMTECH *1980*, 762;
Calderon, N., Lawrence, J. P., Ofstead, E. A.: Adv. Organomet. Chem. *17*, 449 (1979);
Haines, R. J., Leigh, G. J.: Chem. Soc. Rev. *4*, 155 (1975);
Ivin, K. J.: Olefin Metathesis, Academic Press, New York 1983
127. Gilb, W., Schröder, G.: Chem. Ber. *115*, 240 (1982)

128. Röttele, H., Schröder, G.: ibid. *115*, 248 (1982)
129. Lange, G. L., Huggins, M.-A., Neidert, E.: Tetrahedron Lett. *1976*, 4409
130. Lange, G. L., McCarthy, F. C.: ibid. *1978*, 4749
131. Wilson, S. R., Phillips, L. R., Pelister, Y., Huffman, J. C.: J. Am. Chem. Soc. *101*, 7373 (1979)
132. Williams, J. R., Callahan, J. F.: J. Org. Chem. *45*, 4479 (1980)
133. Wender, P. A., Lechleiter, J. C.: J. Am. Chem. Soc. *99*, 267 (1977)
134. Wender, P. A., Lechleiter, J. C.: ibid. *102*, 6340 (1980)
135. Wender, P. A., Hubbs, J. C.: J. Org. Chem. *45*, 365 (1980)
136. Wender, P. A., Letendre, L. J.: ibid. *45*, 367 (1980)
137. Williams, J. R., Callahan, J. F.: J. Chem. Soc. Chem. Commun. *1979*, 404
138. Williams, J. R., Callahan, J. F.: ibid. *1979*, 405
139. Williams, J. R., Callahan, J. F.: J. Org. Chem. *45*, 4475 (1980)
140. Mehta, G., Reddy, D. S., Murty, A. N.: J. Chem. Soc. Chem. Commun. *1983*, 824
141. Scott, L. T., Kirms, M. A., Günther, H., von Puttkamer, H.: J. Am. Chem. Soc. *105*, 1372 (1983)
142. Cohen, T., Bhupathy, M., Matz, J. R.: ibid. *105*, 520 (1983)
143. Naito, T., Nakayama, N., Kaneko, C.: Chem. Lett. *1981*, 423
144. Naito, T., Kaneko, C.: J. Syn. Org. Chem. (Japan) *42*, 51 (1984)
145. Do Khac Manh Duc, Fetizon, M., Hanna, L., Lazare, S.: Synthesis *1981*, 139
146. Wilson, S. R., Mao, D. T.: J. Chem. Soc. Chem. Commun. *1978*, 479
147. Berson, J. A.: Acc. Chem. Res. *1*, 152 (1968)
148. Duke, F. R.: J. Am. Chem. Soc. *69*, 3054 (1947); Bunton, C. A. in Wiberg, K. B. (Ed.): Oxidation in Organic Chemistry, Part A, Academic Press, New York 1965, pp. 367–407
149. Williams, J. R., Caggiano, T. J.: Synthesis *1980*, 1024
150. Van Audenhove, M., De Keukeleire, D., Vandewalle, M.: Tetrahedron Lett. *21*, 1979 (1980)
151. Van Hijfte, L., Vandewalle, M.: ibid. *23*, 2229 (1982)
152. Termont, D., Declercq, P., De Keukeleire, D., Vandewalle, M.: Synthesis *1977*, 46
153. Declercq, P., Vandewalle, M.: J. Org. Chem. *42*, 3447 (1977)
154. Devreese, A. A., Demuynck, M., Declercq, P., Vandewalle, M.: Tetrahedron *39*, 3039 (1983)
155. Devreese, A. A., Demuynck, M., Declercq, P., Vandewalle, M.: ibid. *39*, 3049 (1983)
156. Weinreb, S. M., Cvetovich, R. J.: Tetrahedron ett. *1972*, 1233
157. Liu, H.-J.: Can. J. Chem. *54*, 3113 (1976)
158. Hunter, N. R., MacAlpine, G. A., Liu, H.-J., Valenta, Z.: ibid. *48*, 1436 (1970)
159. Liu, H.-J., Yao, P. C.-L.: ibid. *55*, 822 (1977)
160. Koft, E. R., Smith, III, A. B.: J. Am. Chem. Soc. *104*, 5568 (1982)
161. Boschelli, D., Smith, III, A. B.: Tetrahedron Lett. *22*, 3733 (1981)
162. Smith, III, A. B., Boschelli, D.: J. Org. Chem. *48*, 1217 (1983)
163. Greene, A. E., Teixeira, M. A., Barreiro, E., Cruz, A., Crabbé, P.: ibid. *47*, 2553 (1982)
164. Oppolzer, W., Gorrichon, L., Bird, T. G. C.: Helv. Chim. Acta *64*, 186 (1981)
165. Oppolzer, W., Zutterman, F., Bättig, K.: ibid. *66*, 522 (1983)
166. Baker, W. R., Senter, P. D., Coates, R. M.: J. Chem. Soc. Chem. Commun. *1980*, 1011 Coates, R. M., Senter, P. D., Baker, W. R.: J. Org. Chem. *47*, 3597 (1982)
167. Bagli, J. F., Bogri, T.: Tetrahedron Lett. *1969*, 1639
168. Bagli, J. F., Bogri, T.: J. Org. Chem. *37*, 2132 (1972)
169. Hey, H.: Angew. Chem. *83*, 144 (1971); Angew. Chem. Int. Ed. Engl. *10*, 132 (1971)
170. Wender, P. A., Lechleiter, J. C.: J. Am. Chem. Soc. *100*, 4321 (1978)
171. Ohfune, Y., Misumi, S., Furusaki, A., Shirahama, H., Matsumoto, T.: Tetrahedron Lett. *1977*, 279
172. Krepski, L. R., Hassner, A.: J. Org. Chem. *43*, 3173 (1978)
173. Baldwin, S. W., Landmesser, N. G.: Tetrahedron Lett. *23*, 4443 (1982)
174. Baldwin, S. W., Blomquist, Jr., H. R.: J. Am. Chem. Soc. *104*, 4990 (1982)
175. Johnson, C. R., Barbachyn, M. R.: ibid. *104*, 4290 (1982)
176. Aratani, T., Yoneyoshi, Y., Nagase, T.: Tetrahedron Lett. *1975*, 1707
177. Nozaki, H., Moriati, S., Takaya, H., Noyori, R.: Tetrahedron *24*, 3655 (1968)

178. Nakamura, A.: Pure and Applied Chem. *50*, 37 (1978)
179. Davison, A., Krusell, W. C., Michaelson, R. C.: J. Organomet. Chem. *72*, C7 (1974)
180. Cooke, M. D., Fischer, E. O.: ibid. *56*, 279 (1973)
181. Matsuda, H., Kanai, H.: Chem. Lett. *1981*, 395
182. DeVos, M. J., Krief, A.: Tetrahedron Lett. *24*, 103 (1983)
183. Houge, C., Frisque-Hesbain, A. M., Mockel, A., Ghosez, L., Declercq, J. P., Germain, G., Van Meersche, M.: J. Am. Chem. Soc. *104*, 2920 (1982);
Saimoto, H., Houge, C., Hesbain-Frisque, A.-M., Mockel, A., Ghosez, L.: Tetrahedron Lett. *24*, 2251 (1983);
Bruneel, K., De Keukeleire, D., Vandewalle, M.: J. Chem. Soc. Perkin Trans. 1 *1984*, 1697;
Meyers, A. I., Fleming, S. A.: J. Am. Chem. Soc. *108*, 306 (1986);
For a review, see Paquette, L. A. in Morrison, J. D. (Ed.): Asymmetric synthesis, Volume 3, Academic Press, New York, 1984

Author Index Volumes 101–133

Contents of Vols. 50–100 see Vol. 100
Author and Subject Index Vols. 26–50 see Vol. 50

The volume numbers are printed in italics